Game Theory in Action

Game Theory in Action

An Introduction to Classical and Evolutionary Models

Stephen Schecter
Herbert Gintis

Princeton University Press
Princeton and Oxford

Published by Princeton University Press,
41 William Street, Princeton, New Jersey 08540

In the United Kingdom: Princeton University Press,
6 Oxford Street, Woodstock, Oxfordshire OX20 1TW

press.princeton.edu

ISBN 978-0-691-16764-0
ISBN (pbk.) 978-0-691-16765-7
Library of Congress Control Number: 2015956193

British Library Cataloging-in-Publication Data is available

This book has been composed in Lucida using TeX

Typeset by T&T Productions Ltd, London

Printed on acid-free paper ∞

Printed in the United States of America

1 3 5 7 9 10 8 6 4 2

For Nancy, Erica, and Jeff.
   S.S.
To the math teachers who inspired me:
  Mr. Getty in junior high school,
  Pincus Schub and Walter Gottschalk at the University of Pennsylvania,
  and Lynn Loomis and Oscar Zariski at Harvard University.
   H.G.

Contents

Preface and acknowledgments

Game theory deals with situations in which your payoff depends not only on your own choices but also on the choices of others. How are you supposed to decide what to do, since you cannot control what others will do?

In calculus you learn to maximize and minimize functions, for example to find the cheapest way to build something. This field of mathematics is called optimization. Game theory differs from optimization in that in optimization problems, your payoff depends only on your own choices.

Like the field of optimization, game theory is defined by the problems it deals with, not by the mathematical techniques that are used to deal with them. The techniques are whatever works best.

Also, like the field of optimization, the problems of game theory come from many different areas of study. It is nevertheless helpful to treat game theory as a single mathematical field, since then techniques developed for problems in one area, for example evolutionary biology, become available to another, for example economics.

Game theory has three uses:

(1) Understand the world. For example, game theory helps explain why animals sometimes fight over territory and sometimes don't.
(2) Respond to the world. For example, game theory has been used to develop strategies to win at poker.
(3) Change the world. Often the world is the way it is because people are responding to the rules of a game. Changing the game can change how they act. For example, rules on using energy can be designed to encourage conservation and innovation.

The idea behind the organization of this book is to learn an idea, and then try to use it in as many interesting ways as possible. Because of this organization, the most important idea in game theory, the Nash equilibrium, does not make an appearance until Chapter 3. Two ideas that are more basic—backward induction for games in extensive form, and elimination of dominated strategies for games in normal form—are treated first.

Traditionally, game theory has been viewed as a way to find rational answers to dilemmas. However, since the 1970s biologists have applied game theory to animal behavior, without assuming that animals make rational analyses. Instead they assume that predominant strategies emerge over time as more successful strategies replace less successful ones. This point of view on game theory is now called *evolutionary game theory*. Once one thinks of strategies as changing over time, the mathematical field of differential equations becomes relevant. Because students do not always have a good background in differential equations, we have included an introduction to the area in Chapter 9.

This text grew out of Herb's book [4], which is a problem-centered introduction to modeling strategic interaction. Steve began using Herb's book in 2005 to teach a game theory course in the North Carolina State University Mathematics Department. The course was aimed at upper division mathematics majors and other interested students with some mathematical background (calculus, including some differential equations). Over the following years Steve produced a set of class notes to supplement [4], which was superseded in 2009 by [5]. This text combines material from those two books by Herb and from his recent book [6] with Steve's notes, and it adds some new material.

Examples and problems are the heart of the book. There are more examples in the text than can reasonably be covered in a semester. The instructor can follow her own taste in what to omit.

We suggest, however, that an instructor cover at least the statements of most of the general results. The only general sections that Steve usually omits in his course are on the Folk Theorem, general symmetries of games, and epistemic game theory (Sections 6.8, 7.4, and 8.2, respectively).

The chapters are arranged so that each uses material from the preceding ones; thus the chapters should be covered in order. The one exception is Chapter 6, "More about games in extensive form with complete information": it cannot be covered before Chapter 5, but later chapters do not depend on it, so it could be skipped.

The text includes proofs of general results, written in a fairly typical mathematical style. Steve usually covers just a few of these in his course, since the course is open to students with a limited mathematical background. However, mathematics students who have had previous proof-oriented courses should be able to handle them.

We thank Seth Ditchik and Peter Dougherty at Princeton University Press, who provided faith and wise guidance in our quest to produce a truly distinctive textbook on game theory. We also thank two anonymous referees, whose insightful comments led to many improvements in the book.

We thank Steve's wife Nancy Schecter, whose idea it was to expand a set of class notes into a book, and who was a source of constant encouragement throughout the writing. We also thank the North Carolina State students in ten game theory classes since 2005, and Herb's students around the world, who contributed in countless ways to our understanding of game theory.

Game Theory in Action

Chapter 1

Backward induction

This chapter deals with situations in which two or more opponents take actions one after the other. If you are involved in such a situation, you can try to think ahead to how your opponent might respond to each of your possible actions, bearing in mind that he is trying to achieve his own objectives, not yours. However, we shall see that it may not be helpful to carry this idea too far.

1.1 Tony's Accident

When one of us (Steve) was a college student, his friend Tony caused a minor traffic accident. We'll let Steve tell the story:

The car of the victim, whom I'll call Vic, was slightly scraped. Tony didn't want to tell his insurance company. The next morning, Tony and I went with Vic to visit some body shops. The upshot was that the repair would cost $80.

Tony and I had lunch with a bottle of wine, and thought over the situation. Vic's car was far from new and had accumulated many scrapes. Repairing the few that Tony had caused would improve the car's appearance only a little. We figured that if Tony sent Vic a check for $80, Vic would probably just pocket it. Perhaps, we thought, Tony should ask to see a receipt showing that the repairs had actually been performed before he sent Vic the $80.

A game theorist would represent this situation by the game tree in Figure 1.1. For definiteness, we'll assume that the value to Vic of repairing the damage is $20.

Explanation of the game tree:

(1) Tony goes first. He has a choice of two actions: send Vic a check for $80, or demand a receipt proving that the work has been done.
(2) If Tony sends a check, the game ends. Tony is out $80; Vic will no doubt keep the money, so he has gained $80. We represent these payoffs by the ordered pair $(-80, 80)$; the first number is Tony's payoff, the second is Vic's.

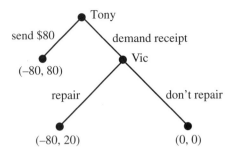

Figure 1.1. Tony's Accident. Tony's payoff is given first.

(3) If Tony demands a receipt, Vic has a choice of two actions: repair the car and send Tony the receipt, or just forget the whole thing.

(4) If Vic repairs the car and sends Tony the receipt, the game ends. Tony sends Vic a check for $80, so he is out $80; Vic uses the check to pay for the repair, so his gain is $20, the value of the repair.

(5) If Vic decides to forget the whole thing, he and Tony each end up with a gain of 0.

Assuming that we have correctly sized up the situation, we see that if Tony demands a receipt, Vic will have to decide between two actions, one that gives him a payoff of $20 and one that gives him a payoff of 0. Vic will presumably choose to repair the car, which gives him a better payoff. Tony will then be out $80.

Our conclusion was that Tony was out $80 whatever he did. We did not like this game.

When the bottle of wine was nearly finished, we thought of a third course of action that Tony could take: send Vic a check for $40, and tell Vic that he would send the rest when Vic provided a receipt showing that the work had actually been done. The game tree now became the one in Figure 1.2.

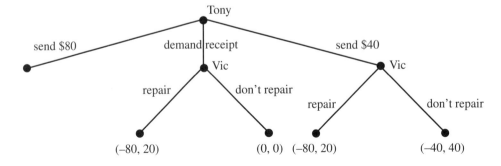

Figure 1.2. Tony's Accident: second game tree. Tony's payoff is given first.

Most of the new game tree looks like the first one. However:

(1) If Tony takes his new action, sending Vic a check for $40 and asking for a receipt, Vic will have a choice of two actions: repair the car, or don't.

(2) If Vic repairs the car, the game ends. Vic will send Tony a receipt, and Tony will send Vic a second check for $40. Tony will be out $80. Vic will use both checks to pay for the repair, so he will have a net gain of $20, the value of the repair.

(3) If Vic does not repair the car, and just pockets the the $40, the game ends. Tony is out $40, and Vic has gained $40.

Again assuming that we have correctly sized up the situation, we see that if Tony sends Vic a check for $40 and asks for a receipt, Vic's best course of action is to keep the money and not make the repair. Thus Tony is out only $40.

Tony sent Vic a check for $40, told him he'd send the rest when he saw a receipt, and never heard from Vic again.

1.2 Games in extensive form with complete information

Tony's Accident is the kind of situation that is studied in game theory, because:

(1) It involves more than one individual.
(2) Each individual has several possible actions.
(3) Once each individual has chosen his actions, payoffs to all individuals are determined.
(4) Each individual is trying to maximize his own payoff.

The key point is that the payoff to an individual depends not only on his own choices, but also on the choices of others.

We gave two models for Tony's Accident, which differed in the sets of actions available to Tony and Vic. Each model was a *game in extensive form with complete information.*

A game in extensive form with complete information consists, to begin with, of the following:

(1) A set *P* of *players.* In Figure 1.2, the players are Tony and Vic.
(2) A set *N* of *nodes.* In Figure 1.2, the nodes are the little black circles. There are eight of them in this case.
(3) A set *B* of *actions* or *moves.* In Figure 1.2, the moves are the lines (seven in this case). Each move connects two nodes, one its *start* and the other its

end. In Figure 1.2, the start of a move is the node at the top of the move, and the end of a move is the node at the bottom of the move.

A *root node* is a node that is not the end of any move. In Figure 1.2, the top node is the only root node.

A *terminal node* is a node that is not the start of any move. In Figure 1.2 there are five terminal nodes.

A *path* is sequence of moves such that the end node of any move in the sequence is the start node of the next move in the sequence. A path is *complete* if it is not part of any longer path. Paths are sometimes called *histories*, and complete paths are called *complete histories.* If a complete path has finite length, it must start at a root node and end at a terminal node.

A game in extensive form with complete information also has:

(4) A function from the set of nonterminal nodes to the set of players. This function, called a *labeling* of the set of nonterminal nodes, tells us which player chooses a move at that node. In Figure 1.2, there are three nonterminal nodes. One is labeled "Tony" and two are labeled "Vic."

(5) For each player, a *payoff function* from the set of complete paths into the real numbers. Usually the players are numbered from 1 to n, and the ith player's payoff function is denoted π_i.

A game in extensive form with complete information is required to satisfy the following conditions:

(a) There is exactly one root node.
(b) If c is any node other than the root node, there is exactly one path from the root node to c.

One way of thinking of (b) is that if you know the node you are at, you know exactly how you got there.

Here are two consequences of assumption (b):

1. Each node other than the root node is the end of exactly one move. (Proof: Let c be a node that is not the root node. It is the end of *at least* one move, because there is a path from the root node to c. If c were the end of two moves m_1 and m_2, then there would be two paths from the root node to c: one from the root node to the start of m_1, followed by m_1; the other from the root node to the start of m_2, followed by m_2. But this can't happen because of assumption (b).)

2. *Every* complete path, not just those of finite length, starts at the root node. (If c is any node other than the root node, there is exactly one path p from the root node to c. If a path that contains c is complete, it must contain p.)

A *finite horizon game* is one in which there is a number K such that every complete path has length at most K. In Chapters 1 to 5 of this book we only discuss finite horizon games. An *infinite horizon game* is one with arbitrarily long paths. We discuss these games in Chapter 6.

In a finite horizon game, the complete paths are in one-to-one correspondence with the terminal nodes. Therefore, in a finite horizon game we can define a player's payoff function by assigning a number to each terminal node.

In Figure 1.2, Tony is Player 1 and Vic is Player 2. Thus each terminal node e has associated with it two numbers, Tony's payoff $\pi_1(e)$ and Vic's payoff $\pi_2(e)$. In Figure 1.2 we have labeled each terminal node with the ordered pair of payoffs $(\pi_1(e), \pi_2(e))$.

A game in extensive form with complete information is *finite* if the number of nodes is finite. (It follows that the number of moves is finite. In fact, the number of moves in a finite game is always one less than the number of nodes.) Such a game is necessarily a finite horizon game.

Games in extensive form with complete information are good models of situations in which players act one after the other, players understand the situation completely, and nothing depends on chance. In Tony's Accident it was important that Tony knew Vic's payoffs, at least approximately, or he would not have been able to choose what to do.

1.3   Strategies

In game theory, a player's *strategy* is a plan for what action to take in every situation that the player might encounter. *For a game in extensive form with complete information, the phrase "every situation that the player might encounter" is interpreted to mean every node that is labeled with his name.*

In Figure 1.2, only one node, the root, is labeled "Tony." Tony has three possible strategies, corresponding to the three actions he could choose at the start of the game. We will call Tony's strategies s_1 (send \$80), s_2 (demand a receipt before sending anything), and s_3 (send \$40).

In Figure 1.2, there are two nodes labeled "Vic." Vic has four possible strategies, which we label $t_1, \ldots, t_4$:

Vic's strategy	If Tony demands receipt	If Tony sends \$40
t_1	repair	repair
t_2	repair	don't repair
t_3	don't repair	repair
t_4	don't repair	don't repair

In general, suppose there are k nodes labeled with a player's name, and there are n_1 possible moves at the first node, n_2 possible moves at the second node, ..., and n_k possible moves at the kth node. A strategy for that player consists of a choice of one of his n_1 moves at the first node, one of his n_2 moves at the second node, ..., and one of his n_k moves at the kth node. Thus the number of strategies available to the player is the product $n_1 n_2 \cdots n_k$.

If we know each player's strategy, then we know the complete path through the game tree, so we know both players' payoffs. With some abuse of notation, we denote the payoffs to Players 1 and 2 when Player 1 uses the strategy s_i and Player 2 uses the strategy t_j by $\pi_1(s_i, t_j)$ and $\pi_2(s_i, t_j)$. For example, $(\pi_1(s_3, t_2), \pi_2(s_3, t_2)) = (-40, 40)$. Of course, in Figure 1.2, this is the pair of payoffs associated with the terminal node on the corresponding path through the game tree.

Recall that if you know the node you are at, you know how you got there. Thus a strategy can be thought of as a plan for how to act after each course the game might take (that ends at a node where it is your turn to act).

1.4 Backward induction

Game theorists often assume that players are *rational*. For a game in extensive form with complete information, rationality is usually considered to imply the following:

- Suppose a player has a choice that includes two moves m and m', and m yields a higher payoff to that player than m'. Then the player will not choose m'.

Thus, if you assume that your opponent is rational in this sense, you must assume that whatever you do, your opponent will respond by doing what is best for him, not what you might want him to do. (Game theory discourages wishful thinking.) Your opponent's response will affect your own payoff. You should therefore take your opponent's likely response into account in deciding on your own action. This is exactly what Tony did when he decided to send Vic a check for $40.

The assumption of rationality motivates the following procedure for selecting strategies for all players in a finite game in extensive form with complete information. This procedure is called *backward induction* or *pruning the game tree*.

(1) Select a node c such that all the moves available at c have ends that are terminal. (Since the game is finite, there must be such a node.)

(2) Suppose Player i is to choose at node c. Among all the moves available to him at that node, find the move m whose end e gives the greatest payoff to Player i. *In the rest of this chapter, and until Chapter 6, we shall only deal with situations in which this move is unique.*

(3) Assume that at node c, Player i will choose the move m. Record this choice as part of Player i's strategy.

(4) Delete from the game tree all moves that start at c. The node c is now a terminal node. Assign to it the payoffs that were previously assigned to the node e.

(5) The game tree now has fewer nodes. If it has just one node, stop. If it has more than one node, return to step 1.

In step 2 we find the move that Player i presumably will make should the course of the game arrive at node c. In step 3 we assume that Player i will in fact make this move, and record this choice as part of Player i's strategy. In step 4 we assign to node c the payoffs to all players that result from this choice, and we "prune the game tree." This helps us take this choice into account when finding the moves players should presumably make at earlier nodes.

In Figure 1.2, there are two nodes for which all available moves have terminal ends: the two where Vic is to choose. At the first of these nodes, Vic's best move is repair, which gives payoffs of $(-80, 20)$. At the second, Vic's best move is don't repair, which gives payoffs of $(-40, 40)$. Thus after two steps of the backward induction procedure, we have recorded the strategy t_2 for Vic, and we arrive at the pruned game tree of Figure 1.3.

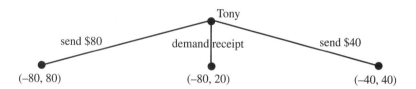

Figure 1.3. Tony's Accident: pruned game tree.

Now the node labeled "Tony" has all its ends terminal. Tony's best move is to send \$40, which gives him a payoff of -40. Thus Tony's strategy is s_3. We delete all moves that start at the node labeled "Tony" and label that node with the payoffs $(-40, 40)$. That is now the only remaining node, so we stop.

Thus the backward induction procedure selects strategy s_3 for Tony and strategy t_2 for Vic, and predicts that the game will end with the payoffs $(-40, 40)$. This is how the game ended in reality.

When you are doing problems using backward induction, you may find that recording parts of strategies and then pruning and redrawing game trees is too slow. Here is another way to do problems. First, find the nodes c such that all moves available at c have ends that are terminal. At each of these nodes, cross out all moves that do not produce the greatest payoff for the player who chooses. If we do this for the game pictured in Figure 1.2, we get Figure 1.4.

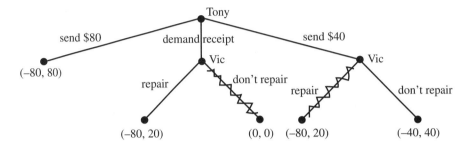

Figure 1.4. Tony's Accident: start of backward induction.

Now you can back up a step. In Figure 1.4 we now see that Tony's three possible moves will produce payoffs to him of -80, -80, and -40. Cross out the two moves that produce payoffs of -80. We obtain Figure 1.5.

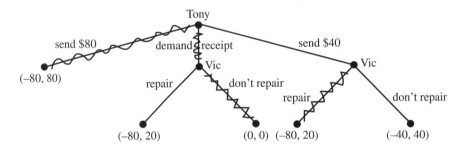

Figure 1.5. Tony's Accident: completion of backward induction.

From Figure 1.5 we can read off each player's strategy; for example, we can see what Vic will do at each of the nodes where he chooses, should that node be reached. We can also see how the game will play out if each player uses the strategy we have found.

In more complicated examples, of course, this procedure will have to be continued for more steps.

The backward induction procedure can fail if, at any point, step 2 produces two moves that give the same highest payoff to the player who is to choose. Figure 1.6 shows an example where backward induction fails. At the node where Player 2 chooses, both available moves give him a payoff of 1. Player 2

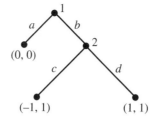

Figure 1.6. Failure of backward induction. As is standard when the players are numbered 1 and 2, Player 1's payoff is given first.

is indifferent between these moves. Hence Player 1 does not know which move Player 2 will choose if Player 1 chooses *b*. Now Player 1 cannot choose between his moves *a* and *b*, since which is better for him depends on which choice Player 2 would make if Player 1 chose *b*.

We return to this issue in Chapter 6.

1.5 Big Monkey and Little Monkey 1

Big Monkey and Little Monkey eat coconuts, which dangle from a branch of the coconut palm. One of them (at least) must climb the tree and shake down the fruit. Then both can eat it. The monkey that doesn't climb will have a head start eating the fruit.

If Big Monkey climbs the tree, he incurs an energy cost of 2 kilocalories (Kc). If Little Monkey climbs the tree, he incurs a negligible energy cost (because he's so little).

A coconut can supply the monkeys with 10 Kc of energy. It will be divided between the monkeys as follows:

	Big Monkey eats	Little Monkey eats
If Big Monkey climbs	6 Kc	4 Kc
If both monkeys climb	7 Kc	3 Kc
If Little Monkey climbs	9 Kc	1 Kc

Let's assume that Big Monkey must decide what to do first. Payoffs are net gains in kilocalories. The game tree is shown in Figure 1.7. Backward induction produces the following strategies:

(1) Little Monkey: If Big Monkey waits, climb. If Big Monkey climbs, wait.
(2) Big Monkey: Wait.

Thus Big Monkey waits. Little Monkey, having no better option at this point, climbs the tree and shakes down the fruit. He scampers quickly down, but to no avail: Big Monkey has gobbled most of the fruit. Big Monkey has a net gain of 9 Kc, Little Monkey 1 Kc.

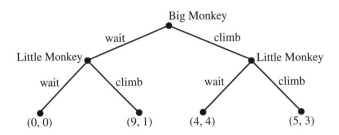

Figure 1.7. Big Monkey and Little Monkey. Big Monkey's payoff is given first.

1.6 Threats, promises, commitments

The game of Big Monkey and Little Monkey has the following peculiarity. Suppose Little Monkey adopts the strategy wait no matter what Big Monkey does. If Big Monkey is convinced that this is in fact Little Monkey's strategy, he sees that his own payoff will be 0 if he waits and 4 if he climbs. His best option is therefore to climb. The payoffs are 4 Kc to each monkey.

Little Monkey's strategy of waiting no matter what Big Monkey does is not "rational" in the sense of the last section, since it involves taking an inferior action should Big Monkey wait. Nevertheless it produces a better outcome for Little Monkey than his "rational" strategy.

A commitment by Little Monkey to wait if Big Monkey waits is called a *threat.* If in fact Little Monkey waits after Big Monkey waits, Big Monkey's payoff is reduced from 9 to 0. Of course, Little Monkey's payoff is also reduced, from 1 to 0. The value of the threat, if it can be made believable, is that it should induce Big Monkey *not* to wait, so that the threat will not have to be carried out.

The ordinary use of the word "threat" includes the idea that the threat, if carried out, would be bad *both* for the opponent *and* for the individual making the threat. Think, for example, of a parent threatening to punish a child, or a country threatening to go to war. If an action would be bad for your opponent and good for you, there is no need to threaten to do it; it is your normal course.

The difficulty with threats is how to make them believable, since if the time comes to carry out the threat, the player making the threat will not want to do it. Some sort of advance commitment is necessary to make the threat believable. Perhaps Little Monkey should break his own leg and show up on crutches!

In this example the threat by Little Monkey works to his advantage. If Little Monkey can somehow convince Big Monkey that he will wait if Big Monkey waits, then from Big Monkey's point of view, the game tree changes to the one shown in Figure 1.8.

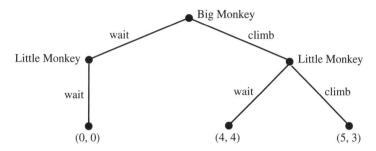

Figure 1.8. Big Monkey and Little Monkey after Little Monkey commits to wait if Big Monkey waits. Big Monkey's payoff is given first.

If Big Monkey uses backward induction on the new game tree, he will climb!

Closely related to threats are *promises*. In the game of Big Monkey and Little Monkey, Little Monkey could make a promise at the node after Big Monkey climbs. Little Monkey could promise to climb. This would increase Big Monkey's payoff at that node from 4 to 5, while decreasing Little Monkey's payoff from 4 to 3. Here, however, even if Big Monkey believes Little Monkey's promise, it will not affect his action in the larger game. He will still wait, getting a payoff of 9.

The ordinary use of the word "promise" includes the idea that it is both good for the other person and bad for the person making the promise. If an action is also good for you, then there is no need to promise to do it; it is your normal course. Like threats, promises usually require some sort of advance commitment to make them believable.

Let us consider threats and promises more generally. Consider a two-player game in extensive form with complete information G. We first consider a node c such that all moves that start at c have terminal ends. Suppose for simplicity that Player 1 is to move at node c. Suppose Player 1's "rational" choice at node c, the one she would make if she were using backward induction, is a move m that gives the two players payoffs (π_1, π_2). Now imagine that Player 1 commits herself to a different move m' at node c, which gives the two players payoffs (π_1', π_2'). If m was the unique choice that gave Player 1 her best payoff, we necessarily have $\pi_1' < \pi_1$, that is, the new move gives Player 1 a lower payoff.

- If $\pi_2' < \pi_2$ (i.e., if the choice m' reduces Player 2's payoff as well), Player 1's commitment to m' at node c is a threat.
- If $\pi_2' > \pi_2$ (i.e., if the choice m' increases Player 2's payoff), Player 1's commitment to m' at node c is a promise.

Now consider any node c where, for simplicity, Player 1 is to move. Suppose Player 1's "rational" choice at node c, the one she would make if she were using backward induction, is a move m. Suppose that if we use backward induction, when we have reduced to a game in which the node c is terminal, the payoffs to the two players at c are (π_1, π_2). Now imagine that Player 1 commits herself to a different move m' at node c. Remove from the game G all other moves that start at c, and all parts of the tree that are no longer connected to the root node once these moves are removed. Call the new game G'. Suppose that if we use backward induction in G', when we have reduced to a game in which the node c is terminal, the payoffs to the two players at c are (π_1', π_2'). Under the uniqueness assumption we have been using, we necessarily have $\pi_1' < \pi_1$:

- If $\pi_2' < \pi_2$, Player 1's commitment to m' at node c is a threat.
- If $\pi_2' > \pi_2$, Player 1's commitment to m' at node c is a promise.

1.7 Ultimatum Game

Player 1 is given 100 one dollar bills. She must offer some of them (1 to 99) to Player 2. If Player 2 accepts the offer, she keeps the bills she was offered, and Player 1 keeps the rest. If Player 2 rejects the offer, neither player gets to keep anything.

Let's assume payoffs are dollars gained in the game. Then the game tree is shown in Figure 1.9.

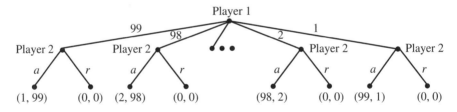

Figure 1.9. Ultimatum Game with dollar payoffs. Player 1 offers a number of dollars to Player 2, then Player 2 accepts (a) or rejects (r) the offer.

Backward induction yields:

- Whatever offer Player 1 makes, Player 2 should accept it, since a gain of even one dollar is better than a gain of nothing.
- Therefore Player 1 should only offer one dollar. That way she gets to keep 99!

However, many experiments have shown that people do no not actually play the Ultimatum Game in accord with this analysis; see the Wikipedia page for this game (http://en.wikipedia.org/wiki/Ultimatum_game). Offers of less than about $40 are typically rejected.

A strategy by Player 2 to reject small offers is an implied threat (actually many implied threats, one for each small offer that she would reject). If Player 1 believes this threat—and experimentation has shown that she should—then she should make a fairly large offer. As in the game of Big Monkey and Little Monkey, a threat to make an "irrational" move, if it is believed, can result in a higher payoff than a strategy of always making the "rational" move.

We should also recognize a difficulty in interpreting game theory experiments. The experimenter can set up an experiment with monetary payoffs, but she cannot ensure that those are the only payoffs that are important to the experimental subject.

In fact, experiments suggest that many people prefer that resources not be divided in a grossly unequal manner, which they perceive as unfair; and that most people are especially concerned when it is they themselves who get the short end of the stick. Thus Player 2 may, for example, feel unhappy about accepting an offer x of less than $50, with the amount of unhappiness equivalent to $4(50 - x)$ dollars (the lower the offer, the greater the unhappiness). Her payoff if she accepts an offer of x dollars is then x if $x > 50$, and $x - 4(50 - x) = 5x - 200$ if $x \leqslant 50$. In this case she should accept offers of greater than $40, reject offers below $40, and be indifferent between accepting and rejecting offers of exactly $40.

Similarly, Player 1 may have payoffs not provided by the experimenter that lead her to make relatively high offers. She may prefer in general that resources not be divided in a grossly unequal manner, even at a monetary cost to herself. Or she may try be the sort of person who does not take advantage of others and may experience a negative payoff when she does not live up to her ideals.

The take-home message is that the payoffs assigned to a player must reflect what is actually important to the player.

We have more to say about the Ultimatum Game in Sections 5.6 and 10.12.

1.8   Rosenthal's Centipede Game

Like the Ultimatum Game, the Centipede Game is a game theory classic.

Mutt and Jeff start with $2 each. Mutt goes first. On a player's turn, he has two possible moves:

(1) Cooperate (*c*): The player does nothing. The game master rewards him with $1.

(2) Defect (*d*): The player steals $2 from the other player.

The game ends when either (1) one of the players defects, or (2) both players have at least $100.

Payoffs are dollars gained in the game. The game tree is shown in Figure 1.10.

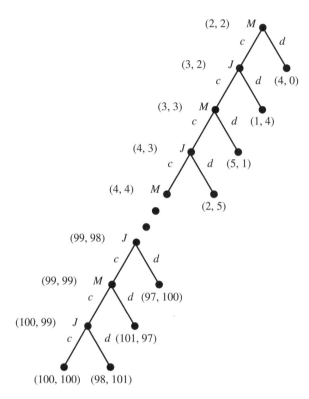

Figure 1.10. Rosenthal's Centipede Game. Mutt's payoff is given first. The amounts the players have accumulated when a node is reached are shown to the left of the node.

A backward induction analysis begins at the only node both of whose moves end in terminal nodes: Jeff's node at which Mutt has accumulated $100 and Jeff has accumulated $99. If Jeff cooperates, he receives $1 from the game master, and the game ends with Jeff having $100. If he defects by stealing $2 from Mutt, the game ends with Jeff having $101. Assuming Jeff is "rational," he will defect. So cross out Jeff's last *c* move in Figure 1.10.

Now we back up a step, to Mutt's last node. We see from the figure that if Mutt cooperates, he will end up with $98, but if he defects, he gets $101 when the game imediately ends. If Mutt is "rational," he will defect. So cross out Mutt's last *c* move.

If we continue the backward induction procedure, we find that it yields the following strategy for each player: whenever it is your turn, defect.

Hence Mutt steals $2 from Jeff at his first turn, and the game ends with Mutt having $4 and Jeff having nothing.

This is a disconcerting conclusion. If you were given the opportunity to play this game, don't you think you could come away with more than $4?

In fact, in experiments, people typically do not defect on the first move. For more information, consult the Wikipedia page for this game, http://en .wikipedia.org/wiki/Centipede_game_(game_theory).

What's wrong with our analysis? Here are a few possibilities.

1. *The players care about aspects of the game other than money.* For example, a player may feel better about himself if he cooperates. Alternatively, a player may want to *seem* cooperative, because this normally brings benefits. If a player wants to be, or to seem, cooperative, we should take account of this desire in assigning his payoffs.

2. *The players use a rule of thumb instead of analyzing the game.* People do not typically make decisions on the basis of a complicated rational analysis. Instead they follow rules of thumb, such as be cooperative and don't steal. In fact, it may not be rational to make most decisions on the basis of a complicated rational analysis, because (a) the cost in time and effort of doing the analysis may be greater than the advantage gained, and (b) if the analysis is complicated enough, you are liable to make a mistake anyway.

3. *The players use a strategy that is correct for a different, more common situation.* We do not typically encounter "games" that we know in advance have exactly or at most n stages, where n is a large number. Instead, we typically encounter games with an unknown number of stages. If the Centipede Game had an unknown number of stages, there would be no place to start a backward induction. In Chapter 6 we will study a class of such games for which it is rational to cooperate as long as your opponent does. When we encounter the unusual situation of a game with at most 196 stages, which is the case with the Centipede Game, perhaps we use a strategy that is correct for the more common situation of a game with an unknown number of stages.

However, the most interesting possibility is that the logical basis for believing that rational players will use long backward inductions is suspect. We address this issue in Section 1.13.

1.9 Continuous games

In the games we have considered so far, when it is a player's turn to move, she has only a finite number of choices. In the remainder of this chapter, we consider some games in which each player may choose an action from an interval of real numbers. For example, if a firm must choose the price to charge for an item, we can imagine that the price could be any nonnegative real number. This allows us to use the power of calculus to find which price produces the best payoff to the firm.

More precisely, we consider games with two players, Player 1 and Player 2. Player 1 goes first. The moves available are all real numbers s in some interval I. Next it is Player 2's turn. The moves available are all real numbers t in some interval J. Player 2 observes Player 1's move s and then chooses her move t. The game is now over, and payoffs $\pi_1(s, t)$ and $\pi_2(s, t)$ are calculated.

Does such a game satisfy the definition that we gave in Section 1.2 of a game in extensive form with complete information? Yes, it does. In the previous paragraph, to describe the type of game we want to consider, we only described the moves, not the nodes. However, the nodes are still there. There is a root node at which Player 1 must choose her move s. Each move s ends at a new node, at which Player 2 must choose t. Each move t ends at a terminal node. The set of all complete paths is the set of all pairs (s, t) with s in I and t in J. Since we described the game in terms of moves, not nodes, it was easier to describe the payoff functions as assigning numbers to complete paths, not as assigning numbers to terminal nodes. That is what we did: $\pi_1(s, t)$ and $\pi_2(s, t)$ assign numbers to each complete path.

Such a game is not finite, but it is a finite horizon game: the length of the longest path is 2.

Let us find strategies for Players 1 and 2 using the *idea* of backward induction. Backward induction as we described it in Section 1.4 cannot be used, because the game is not finite.

We begin with the last move, which is Player 2's. Assuming she is rational, she will observe Player 1's move s and then choose t in J to maximize the function $\pi_2(s, t)$ with s fixed. For fixed s, $\pi_2(s, t)$ is a function of one variable t. Suppose it takes on its maximum value in J at a unique value of t. This number t is Player 2's *best response* to Player 1's move s. Normally the best response t will depend on s, so we write $t = b(s)$. The function $t = b(s)$ gives a strategy for Player 2; that is, it gives Player 2 a choice of action for every possible choice s in I that Player 1 might make.

Player 1 should choose s taking into account Player 2's strategy. If Player 1 assumes that Player 2 is rational and hence will use her best-response strategy, then Player 1 should choose s in I to maximize the function $\pi_1(s, b(s))$. This is again a function of one variable.

1.10 Stackelberg's model of duopoly 1

In a *duopoly*, a certain good is produced by just two firms, which we label 1 and 2. In Stackelberg's model of duopoly (Wikipedia article: http://

en.wikipedia.org/wiki/Stackelberg_duopoly), each firm tries to maximize its own profit by choosing an appropriate level of production. Firm 1 chooses its level of production first; then Firm 2 observes this choice and chooses its own level of production. Would you rather run Firm 1 or Firm 2?

Let s be the quantity produced by Firm 1 and let t be the quantity produced by Firm 2. Then the total quantity of the good produced is $q = s + t$. The market price p of the good depends on q: $p = \phi(q)$. At this price, everything produced can be sold.

Suppose Firm 1's cost to produce the quantity s of the good, which we denote $c_1(s)$, is $4s$, and Firm 2's cost to produce the quantity t of the good, which we denote $c_2(t)$, is $4t$. In other words, both Firm 1 and Firm 2 have the same unit cost of production 4.

1.10.1 First model.

We assume that price falls linearly as total production of the two firms increases. In particular, we assume

$$p = 20 - 2(s + t). \tag{1.1}$$

We denote the profits of the two firms by π_1 and π_2. Now profit is revenue minus cost, and revenue is price times quantity sold. Since the price depends on $q = s + t$, each firm's profit depends in part on how much is produced by the other firm. More precisely,

$$\pi_1(s,t) = ps - c_1(s) = (20 - 2(s + t))s - 4s = (16 - 2t)s - 2s^2, \tag{1.2}$$
$$\pi_2(s,t) = pt - c_2(t) = (20 - 2(s + t))t - 4t = (16 - 2s)t - 2t^2. \tag{1.3}$$

We regard the profits as payoffs in a game. The players are Firms 1 and 2.

In this subsection we allow the levels of production s and t to be any real numbers, even negative numbers and numbers large enough to make the price negative. This doesn't make sense economically, but it avoids mathematical complications.

Since Firm 1 chooses s first, we begin our analysis by finding Firm 2's best response $t = b(s)$. To do this, we must find where the function $\pi_2(s,t)$, with s fixed, has its maximum. Since $\pi_2(s,t)$ with s fixed has a graph that is just an upside-down parabola, we can do this by taking the derivative with respect to t and setting it equal to 0:

$$\frac{\partial \pi_2}{\partial t} = 16 - 2s - 4t = 0.$$

If we solve this equation for t, we will have Firm 2's best-response function:

$$t = b(s) = 4 - \tfrac{1}{2}s.$$

Finally, we must maximize $\pi_1(s, b(s))$, the payoff that Firm 1 can expect from each choice s, assuming that Firm 2 uses its best-response strategy. From (1.2), we have

$$\pi_1(s, b(s)) = \pi_1(s, 4 - \tfrac{1}{2}s) = (16 - 2(4 - \tfrac{1}{2}s))s - 2s^2 = 8s - s^2.$$

Again this function has a graph that is an upside-down parabola, so we can find where it is maximum by taking the derivative and setting it equal to 0:

$$\frac{d}{ds}\pi_1(s, b(s)) = 8 - 2s = 0 \quad \Rightarrow \quad s = 4.$$

Thus $\pi_1(s, b(s))$ is maximum at $s^* = 4$. Given this choice of production level for Firm 1, Firm 2 chooses the production level

$$t^* = b(s^*) = 4 - \tfrac{1}{2}4 = 2.$$

The price is

$$p^* = 20 - 2(s^* + t^*) = 20 - 2(4 + 2) = 8.$$

From (1.2) and (1.3), the profits are

$$\pi_1(s^*, t^*) = (16 - 2 \cdot 2)4 - 2 \cdot 4^2 = 16,$$
$$\pi_2(s^*, t^*) = (16 - 2 \cdot 4)2 - 2 \cdot 2^2 = 8.$$

Firm 1 has twice the level of production and twice the profit of Firm 2. It is better to run the firm that chooses its production level first.

1.10.2 Second model. As remarked, the model in the previous subsection has a disconcerting aspect: the levels of production s and t, and the price p, are all allowed to be negative. We now complicate the model to deal with this objection.

First, we only allow nonnegative production levels: $0 \leqslant s < \infty$ and $0 \leqslant t < \infty$. Second, we assume that if total production rises above 10, the level at which formula (1.1) for the price gives 0, then the price is 0, not the negative number given by formula (1.1):

$$p = \begin{cases} 20 - 2(s + t) & \text{if } s + t < 10, \\ 0 & \text{if } s + t \geqslant 10. \end{cases}$$

We again ask the question, what will be the production level and profit of each firm?

The payoff is again the profit, but the formulas are different:

$$\pi_1(s,t) = ps - c_1(s) = \begin{cases} (20 - 2(s+t))s - 4s & \text{if } 0 \leqslant s+t < 10, \\ -4s & \text{if } s+t \geqslant 10, \end{cases} \tag{1.4}$$

$$\pi_2(s,t) = pt - c_2(t) = \begin{cases} (20 - 2(s+t))t - 4t & \text{if } 0 \leqslant s+t < 10, \\ -4t & \text{if } s+t \geqslant 10. \end{cases} \tag{1.5}$$

We again begin our analysis by finding Firm 2's best response $t = b(s)$.

Unit cost of production is 4. If Firm 1 produces so much that all by itself it drives the price down to 4 or lower, there is no way for Firm 2 to make a positive profit. In this case Firm 2's best response is to produce nothing: that way its profit is 0, which is better than losing money.

Firm 1 by itself drives the price p down to 4 when $20 - 2s = 4$, that is, when its level of production is $s = 8$. We conclude that if Firm 1's level of production s is 8 or higher, Firm 2's best response is 0.

In contrast, if Firm 1 produces $s < 8$, it leaves the price above 4 and gives Firm 2 an opportunity to make a positive profit. In this case Firm 2's profit is given by

$$\pi_2(s,t) = \begin{cases} (20 - 2(s+t))t - 4t = (16 - 2s)t - 2t^2 & \text{if } 0 \leqslant t < 10 - s, \\ -4t & \text{if } t \geqslant 10 - s. \end{cases}$$

See Figure 1.11.

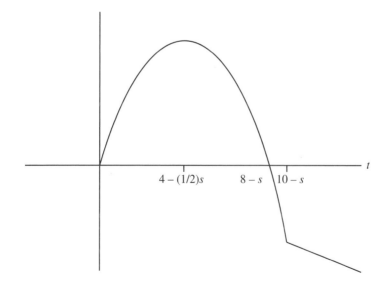

Figure 1.11. Graph of $\pi_2(s,t)$ for fixed $s < 8$ in the numerical example.

From the figure, the function $\pi_2(s,t)$ with s fixed is maximum where $(\partial \pi_2/\partial t)(s,t) = 0$, which occurs at $t = 4 - \frac{1}{2}s$.

Thus Firm 2's best-response function is

$$b(s) = \begin{cases} 4 - \frac{1}{2}s & \text{if } 0 \leqslant s < 8, \\ 0 & \text{if } s \geqslant 8. \end{cases}$$

We now turn to calculating $\pi_1(s, b(s))$, the payoff that Firm 1 can expect from each choice s, assuming that Firm 2 uses its best-response strategy. Notice that for $0 \leqslant s < 8$, we have

$$s + b(s) = s + 4 - \frac{1}{2}s = 4 + \frac{1}{2}s < 4 + \frac{1}{2}8 = 8 < 10.$$

Therefore, for $0 \leqslant s < 8$, we use the first line of (1.4) to calculate $\pi_1(s, b(s))$:

$$\pi_1(s, b(s)) = \pi_1(s, 4 - \tfrac{1}{2}s) = (16 - 2(4 - \tfrac{1}{2}s))s - 2s^2 = 8s - s^2.$$

Firm 1 will not choose an $s \geqslant 8$, since, as we have seen, that would force the price down to the cost of production 4 or lower. Therefore we will not bother to calculate $\pi_1(s, b(s))$ for $s \geqslant 8$.

The function $\pi_1(s, b(s))$ on the interval $0 \leqslant s \leqslant 8$ is maximum at $s^* = 4$, where the derivative of $8s - s^2$ is 0, just as in our first model. The value of $t^* = b(s^*)$ is also the same, as are the price and profits.

1.11 Stackelberg's model of duopoly 2

In this section we give a more general treatment of Stackelberg's model of duopoly.

1.11.1 First model. In this subsection we make the following assumptions, which generalize those used in Subsection 1.10.1.

(1) Price falls linearly with total production. In other words, there are positive numbers α and β such that the formula for the price is

$$p = \alpha - \beta(s + t).$$

(2) Each firm has the same unit cost of production $c > 0$. Thus $c_1(s) = cs$ and $c_2(t) = ct$.

(3) $\alpha > c$. In other words, the price of the good when very little is produced is greater than the unit cost of production. If this assumption is violated, the good will not be produced.

(4) The production levels s and t can be any real numbers.

Firm 1 chooses its level of production s first. Then Firm 2 observes s and chooses t. We ask the question, what will be the production level and profit of each firm?

The payoffs are

$$\pi_1(s,t) = ps - c_1(s) = (\alpha - \beta(s+t))s - cs = (\alpha - \beta t - c)s - \beta s^2,$$
$$\pi_2(s,t) = pt - c_2(t) = (\alpha - \beta(s+t))t - ct = (\alpha - \beta s - c)t - \beta t^2.$$

We find Firm 2's best response $t = b(s)$ by finding where the function $\pi_2(s,t)$, with s fixed, has its maximum. Since $\pi_2(s,t)$ with s fixed has a graph that is an upside-down parabola, we just take the derivative with respect to t and set it equal to 0:

$$\frac{\partial \pi_2}{\partial t} = \alpha - \beta s - c - 2\beta t = 0.$$

We solve this equation for t to find Firm 2's best-response function:

$$t = b(s) = \frac{\alpha - c}{2\beta} - \frac{1}{2}s.$$

Finally, we must maximize $\pi_1(s, b(s))$, the payoff that Firm 1 can expect from each choice s, assuming that Firm 2 uses its best-response strategy. We have

$$\pi_1(s, b(s)) = \pi_1\left(s, \frac{\alpha - c}{2\beta} - \frac{1}{2}s\right) = \left(\alpha - \beta\left(\frac{\alpha - c}{2\beta} - \frac{1}{2}s\right) - c\right)s - \beta s^2$$
$$= \frac{\alpha - c}{2}s - \frac{\beta}{2}s^2.$$

Again this function has a graph that is an upside-down parabola, so we can find where it is maximum by taking the derivative and setting it equal to 0:

$$\frac{d}{ds}\pi_1(s, b(s)) = \frac{\alpha - c}{2} - \beta s = 0 \quad \Rightarrow \quad s = \frac{\alpha - c}{2\beta}.$$

Thus $\pi_1(s, b(s))$ is maximum at $s^* = (\alpha - c)/2\beta$. Given this choice of production level for Firm 1, Firm 2 chooses the production level

$$t^* = b(s^*) = \frac{\alpha - c}{4\beta}.$$

Since we assumed $\alpha > c$, the production levels s^* and t^* are positive, which makes sense. The price is

$$p^* = \alpha - \beta(s^* + t^*) = \alpha - \beta\left(\frac{\alpha - c}{2\beta} + \frac{\alpha - c}{4\beta}\right) = \frac{1}{4}\alpha + \frac{3}{4}c = c + \frac{1}{4}(\alpha - c).$$

Since $\alpha > c$, this price is greater than the cost of production c, which also makes sense.

The profits are

$$\pi_1(s^*, t^*) = \frac{(\alpha - c)^2}{8\beta}, \qquad \pi_2(s^*, t^*) = \frac{(\alpha - c)^2}{16\beta}.$$

As in our numerical example, Firm 1 has twice the level of production and twice the profit of Firm 2.

1.11.2   Second model.   As in Subsection 1.10.2, we now complicate the model to prevent the levels of production s and t and the price p from taking negative values. We replace assumption (1) in Subsection 1.11.1 with the following:

(1′) Price falls linearly with total production until it reaches 0; for higher total production, the price remains 0. In other words, there are positive numbers α and β such that the formula for the price is

$$p = \begin{cases} \alpha - \beta(s + t) & \text{if } s + t < \frac{\alpha}{\beta}, \\ 0 & \text{if } s + t \geqslant \frac{\alpha}{\beta}. \end{cases}$$

Assumptions (2) and (3) remain unchanged. We replace assumption (4) with:

(4′) The production levels s and t must be nonnegative: $0 \leqslant s < \infty$ and $0 \leqslant t < \infty$.

We again ask the question, what will be the production level and profit of each firm?

The payoffs in the general case are

$$\pi_1(s, t) = ps - c_1(s) = \begin{cases} (\alpha - \beta(s + t))s - cs & \text{if } 0 \leqslant s + t < \frac{\alpha}{\beta}, \\ -cs & \text{if } s + t \geqslant \frac{\alpha}{\beta}, \end{cases} \tag{1.6}$$

$$\pi_2(s, t) = pt - c_2(t) = \begin{cases} (\alpha - \beta(s + t))t - ct & \text{if } 0 \leqslant s + t < \frac{\alpha}{\beta}, \\ -ct & \text{if } s + t \geqslant \frac{\alpha}{\beta}. \end{cases} \tag{1.7}$$

As usual we begin our analysis by finding Firm 2's best response $t = b(s)$.

If Firm 1 produces so much that by itself it drives the price down to the unit cost of production c or lower, then Firm 2 canot make a positive profit. In this case Firm 2's best response is to produce nothing. Firm 1 by itself drives the price p down to c when $\alpha - \beta s = c$, that is, when $s = (\alpha - c)/\beta$. We conclude that if $s \geqslant (\alpha - c)/\beta$, Firm 2's best response is 0.

In contrast, if Firm 1 produces $s < (\alpha - c)/\beta$, it leaves the price above c and gives Firm 2 an opportunity to make a positive profit. In this case Firm 2's profit is given by

$$\pi_2(s,t) = \begin{cases} (\alpha - \beta(s+t))t - ct = (\alpha - \beta s - c)t - \beta t^2 & \text{if } 0 \leqslant t < \frac{\alpha}{\beta} - s, \\ -ct & \text{if } t \geqslant \frac{\alpha}{\beta} - s. \end{cases}$$

See Figure 1.12. From the figure, the function $\pi_2(s,t)$ with s fixed is maximum where $(\partial \pi_2/\partial t)(s,t) = 0$, which occurs at

$$t = \frac{\alpha - c}{2\beta} - \frac{1}{2}s.$$

Thus Firm 2's best-response function is:

$$b(s) = \begin{cases} \dfrac{\alpha - c}{2\beta} - \dfrac{1}{2}s & \text{if } 0 \leqslant s < \dfrac{\alpha - c}{\beta}, \\ 0 & \text{if } s \geqslant \dfrac{\alpha - c}{\beta}. \end{cases}$$

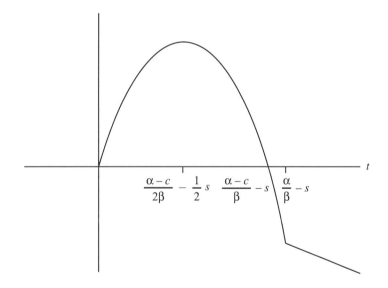

Figure 1.12. Graph of $\pi_2(s,t)$ for fixed $s < (\alpha - c)/\beta$.

We now calculate $\pi_1(s, b(s))$, the payoff that Firm 1 can expect from each choice s, assuming that Firm 2 uses its best-response strategy. Notice that for $0 \leqslant s < (\alpha - c)/\beta$, we have

$$s + b(s) = s + \frac{\alpha - c}{2\beta} - \frac{1}{2}s = \frac{\alpha - c}{2\beta} + \frac{1}{2}s < \frac{\alpha - c}{2\beta} + \frac{\alpha - c}{2\beta} = \frac{\alpha - c}{\beta} < \frac{\alpha}{\beta}.$$

Therefore, for $0 \leqslant s < (\alpha - c)/\beta$, we use the first line of formula (1.6) to calculate $\pi_1(s, b(s))$:

$$\pi_1(s, b(s)) = \pi_1\left(s, \frac{\alpha - c}{2\beta} - \frac{1}{2}s\right) = \left(\alpha - \beta\left(s + \frac{\alpha - c}{2\beta} - \frac{1}{2}s\right)\right)s - cs$$

$$= \frac{\alpha - c}{2}s - \frac{\beta}{2}s^2.$$

Firm 1 will not choose an $s \geqslant (\alpha - c)/\beta$, since, as we have seen, that would force the price down to c or lower. Therefore we will not bother to calculate $\pi_1(s, b(s))$ for $s \geqslant (\alpha - c)/\beta$.

The function $\pi_1(s, b(s))$ on the interval $0 \leqslant s \leqslant (\alpha - c)/\beta$ is maximum at $s^* = (\alpha - c)/2\beta$, where the derivative of $\frac{1}{2}(\alpha - c)s - \frac{1}{2}\beta s^2$ is 0, just as in our first model. The value of $t^* = b(s^*)$ is also the same, as are the price and profits.

1.12 Backward induction for finite horizon games

Backward induction as defined in Section 1.4 does not apply to any game that is not finite. However, a variant of backward induction can be used on any finite horizon game with complete information. It is actually this variant that we have been using since Section 1.9.

Let us describe this variant of backward induction in general. The idea is that, in a game that is not finite, we cannot remove nodes one by one, because we will never finish. Instead we must remove big collections of nodes at each step.

(1) Let $k \geqslant 1$ be the length of the longest path in the game. (This number is finite, since we are dealing with a finite horizon game.) Consider the collection C of all nodes c such that every move that starts at c is the last move in a path of length k. Each such move has an end that is terminal.

(2) For each node c in C, identify the player $i(c)$ who is to choose at node c. Among all the moves available to her at that node, find the move $m(c)$ whose end gives the greatest payoff to Player $i(c)$. We assume that this move is unique.

(3) Assume that at each node c in C, Player $i(c)$ will choose the move $m(c)$. Record this choice as part of Player $i(c)$'s strategy.

(4) Delete from the game tree all moves that start at one of the nodes in C. The nodes c in C are now terminal nodes. Assign to each node c in C the payoffs that were previously assigned to the node at the end of the move $m(c)$.

(5) In the new game tree, the length of the longest path is now $k-1$. If $k-1 = 0$, stop. Otherwise, return to step 1.

1.13 Critique of backward induction

The basic insight of backward induction is that you should think ahead to how your opponent, acting in his own interest, is liable to react to what you do, and act accordingly to maximize your chance of success. This idea clearly makes sense even in situations that are not as completely defined as the games we analyze. For example, the mixed martial arts trainer Greg Jackson has analyzed countless fight videos and used them to make game trees showing what moves lead to what responses. From these game trees he can figure out which moves in various situations will increase the likelihood of a win. As another example, consider the game of chess. Because of the rule that a draw results when a position is repeated three times, the game tree for chess is finite. Unfortunately, it has 10^{123} nodes and hence is far too big for a computer to analyze. (The number of atoms in the observable universe is estimated to be around 10^{80}.) Thus computer chess programs cannot use backward induction from the terminal nodes. Instead they investigate paths through the game tree from a given position to a given depth and assign values to the end nodes based on estimates of the probability of winning from that position. They then use backward induction from those nodes.

Despite successes like these, it is not clear that backward induction is always a good guide to choosing a move.

Let's first consider Tony's Accident. To justify using backward induction at all, Tony has to assume that Vic will always choose his own best move in response to Tony's move. In addition, Tony should know Vic's payoffs, or at least he should know the order in which Vic values the different outcomes, so that he will know which of Vic's available moves Vic will choose in response to Tony's move. If Tony does not know the order in which Vic values the outcomes, he can still use backward induction based on his belief about Vic's order. This is what Tony did. The success of the procedure then depends on the correctness of Tony's beliefs about Vic.

In Chapter 6 we consider a game, the Samaritan's Dilemma, that raises an additional issue. In that game, Daughter wants to decide how much to save from her earnings this year toward her college expenses next year. Father will then observe how much she saves and chip in some of his own earnings. To figure out how much to give, he will balance his desire to keep his earnings to spend on himself against his desire to help his daughter. To justify her use of backward induction in this situation, Daughter has to assume that she knows Father's desires well enough to figure out his best response, from his own point of view, to each of her possible saving levels. She also has to assume that Father will actually make his best response. Here this second assumption becomes hard to justify. To justify it, she needs to assume both that Father is able to figure out his best response, and that he is willing to do so. Recall from our discussion of the Centipede Game that it may not even be rational for Father to use a complicated rational analysis to figure out what to do.

Finally, let's consider the Centipede Game. Would a rational player in the Centipede Game (Section 1.8) really defect at his first opportunity, as is required by backward induction? We examine this question under the assumption that the payoffs in the Centipede Game are exactly as given in Figure 1.10, that both players know these payoffs, and that both players are rational. The assumption that players know the payoffs and are rational motivates backward induction. The issue now is whether the assumption that players know the payoffs and are rational *requires* them to use the moves recommended by backward induction.

By a rational player, we mean one whose preferences are consistent enough to be represented by a payoff function; who attempts to discern the facts about the world; who forms beliefs about the world consistent with the facts he has discerned; and who acts on the basis of his beliefs to best achieve his preferred outcomes.

With this "definition" of a rational player in mind, let us consider the first few steps of backward induction in the Centipede Game.

1. If the node labeled $(100, 99)$ in Figure 1.10 is reached, Jeff will see that if he defects, his payoff is 101, and if he cooperates, his payoff is 100. *Since Jeff is rational*, he defects.

2. If the node labeled $(99, 99)$ in Figure 1.10 is reached, Mutt will see that if he defects, his payoff is 101. If he cooperates, the node labeled $(100, 99)$ is reached. *If Mutt believes that Jeff is rational*, then he sees that Jeff will defect at that node, leaving Mutt with a payoff of only 98. *Since Mutt is rational*, he defects.

3. If the node labeled $(99, 98)$ in Figure 1.10 is reached, Jeff will see that if he defects, his payoff is 100. If he cooperates, the node labeled $(99, 99)$ is reached. *If Jeff believes that Mutt believes that Jeff is rational*, and *if Jeff believes that Mutt is rational*, then Jeff concludes that Mutt will act as described in step 2. This would leave Jeff with a payoff of 97. *Since Jeff is rational*, he defects.

4. You probably see that this is getting complicated fast, but let's do one more step. If the node labeled $(98, 98)$ (not shown in Figure 1.10) is reached, Mutt will see that if he defects, his payoff is 100. If he cooperates, the node labeled $(99, 98)$ is reached. *If Mutt believes that Jeff believes that Mutt believes that Jeff is rational*, and *if Mutt believes that Jeff believes that Mutt is rational*, and *if Mutt believes that Jeff is rational*, then Mutt concludes that Jeff will act as described in step 3. This would leave Mutt with a payoff of 97. *Since Mutt is rational*, he defects.

You can see that by the time we get back to the root node in Figure 1.10, at step 196 in this process, Mutt must hold many complicated beliefs to justify using backward induction to choose his move! (At the kth step in this process, the player who chooses must hold $k - 1$ separate beliefs about the other player, the most complicated of which requires $k - 1$ uses of word "believes" to state.) The question is whether a rational player, "who attempts to discern the facts about the world, who forms beliefs about the world consistent with the facts he has discerned," is *required* to hold such complicated beliefs.

Now suppose that Mutt decides, for whatever reason, to cooperate at his first turn. Then Jeff can conclude that Mutt did not hold the complicated beliefs that would induce him to defect at his first turn. If Jeff believes that at Mutt's second turn, he will hold the (slightly less complicated) beliefs that would induce him to defect then, then Jeff should defect at his own first turn. Clearly, rationality does not require Jeff to believe this. Thus a rational Jeff may well decide to cooperate at his first turn.

In this way, rational players may well cooperate through many stages of the Centipede Game. More generally, *rationality does not require players to follow the strategy dictated by a long backward induction.*

Backward induction *is* required by an assumption called Common Knowledge of Rationality: the players are rational; each player believes that the other players are rational; each player believes that the other players believe that the other players are rational; and so on, for as many steps as are required by the game under discussion. In any particular situation, this assumption may hold, but the assumption that players are rational does not by itself imply Common Knowledge of Rationality.

We shall return to the question of players' beliefs in Section 8.2.

1.14 Problems

1.14.1 Congress vs. the President. Congress is working on a homeland security spending bill. Congress wants the bill to include $10 million for each member's district for "important projects." The President wants the bill to include $100 million to upgrade her airplane, Air Force One, to the latest model, Air Force One Extreme. The President can sign or veto any bill that Congress passes.

Congress's payoffs are 1 if it passes a bill (voters like to see Congress take action), 2 if the President signs a bill that contains money for important projects, and -1 if the President signs a bill that contains money for Air Force One Extreme. Payoffs are added; for example, if the President signs a bill that contains money for both, Congress's total payoff is $1 + 2 + (-1) = 2$.

The President's payoffs are -1 if she signs a bill that contains money for important projects, and 2 if she signs a bill that contains money for Air Force One Extreme. Her payoffs are also added.

The game tree in Figure 1.13 illustrates the situation. In the tree, $C =$ Congress, $P =$ President; $n =$ no bill passed, $i =$ important projects passed, $a =$ Air Force One Extreme passed, $b =$ both passed; $s =$ President signs the bill, $v =$ President vetoes the bill. The first payoff is Congress's, the second is the President's.

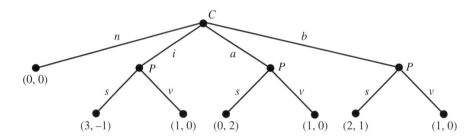

Figure 1.13. Congress vs. the President.

(1) Use backward induction to predict what will happen.
(2) Suppose the Constitution were changed so that the President could veto parts of bills she doesn't like but still sign the rest. Draw the new tree and use backward induction to predict what will happen. (The only change in the game is that if Congress passes a bill containing both important projects and Air Force One Extreme, the President will have four choices: sign, veto, sign important projects but veto Air Force One Extreme, sign Air Force One Extreme but veto important projects.)

1.14.2 Battle of the Chain Stores. Sub Station has the only sub restaurant in Town A and the only sub restaurant in Town B. The sub market in each town yields a profit of $100K per year. Rival Sub Machine is considering opening a restaurant in Town A in year 1. If it does, the two stores will split the profit from the sub market there. However, Sub Machine will have to pay setup costs for its new store. These costs are $25K in a store's first year.

Sub Station fears that if Sub Machine is able to make a profit in Town A, it will open a store in Town B the following year. Sub Station is considering a price war: if Sub Machine opens a store in either town, it will lower prices in that town, forcing Sub Machine to do the same, to the point where profits from the sub market in that town drop to 0.

The game tree in Figure 1.14 is one way to represent the situation. It takes into account net profits from Towns A and B in years 1 and 2, and it assumes that if Sub Machine loses money in A, it will not open a store in B.

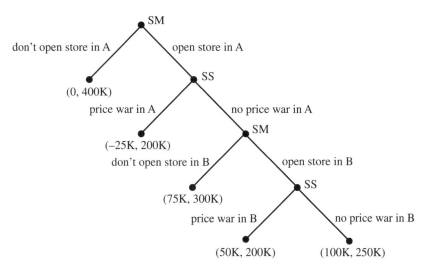

Figure 1.14. SM is Sub Machine, SS is Sub Station. Sub Machine's profits are listed first, Sub Station's profits are second.

The entry $(100K, 250K)$ in the game tree comes about as follows. If there are no price wars, Sub Machine makes net profits of $25K in Town A in year 1, $50K in Town A in year 2, and $25K in Town B in year 2, for a total of $100K. Sub Station makes profits of $50K in Town A in year 1, $50K in Town A in year 2, $100K in Town B in year 1, and $50K in Town B in year 2, for a total of $250K.

(1) Explain the entry $(50K, 200K)$ in the game tree.
(2) Use backward induction to figure out what Sub Machine and Sub Station should do.
(3) How might Sub Station try to obtain a better outcome by using a threat?

1.14.3 Kidnapping. A criminal kidnaps a child and demands a ransom $r > 0$. The value to the parents of the child's return is $v > 0$. If the kidnapper frees the child, he incurs a cost $f > 0$. If he kills the child, he incurs a cost $k > 0$. These costs include the feelings of the kidnapper in each case, the likelihood of being caught in each case, the severity of punishment in each case, and whatever else is relevant. We assume $v > r$. The parents choose first whether to pay the ransom or not, then the kidnapper decides whether to free the child or kill the child. The game tree is shown in Figure 1.15.

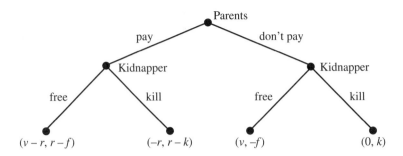

Figure 1.15. Kidnapping. The parents' payoff is given first.

(1) Explain the payoffs.

(2) Suppose $k > f$. Use backward induction to show that the parents should not pay the ransom.

(3) Suppose $f > k$. Use backward induction to show that the parents should not pay the ransom.

(4) Suppose $k > f$. Find a threat the kidnapper can make to try to get the parents to pay the ransom.

(5) Suppose $f > k$. Find a promise the kidnapper can make to try to get the parents to pay the ransom.

(6) Suppose $f > k$, and the parents will incur a guilt cost g if they don't pay the ransom and the child is killed. How large must g be to get the parents to pay the ransom?

1.14.4 The White House Tapes. On March 1, 1974, a grand jury indicted seven former aides to U.S. President Richard Nixon for attempting to cover up White House involvement in a burglary of Democratic National Committee headquarters at the Watergate complex in Washington. On April 18, the judge in the case, John Sirica, issued a subpoena for tapes of President Nixon's conversations with the defendants. The President's attorney, James St. Clair, attempted to delay responding to the subpoena. The prosecutor, Leon Jaworski, then used an unusual procedure to appeal directly to the Supreme Court and request that the Court order the President to supply the tapes. The Court heard oral arguments on July 8, and the justices met on July 9 to decide the case.

One justice, William Rehnquist, withdrew from the case, probably because he had worked in President Nixon's Justice Department. Of the remaining eight justices, six quickly agreed to uphold the prosecutor's request. Two justices, Warren Burger and Harry Blackmun, were reluctant to uphold the prosecutor's request, because they thought his direct appeal to the Supreme Court was improper.

Also on July 9, President Nixon's attorney said that the President had "not yet decided" whether he would supply the tapes if the Supreme Court ordered him to. This statement was probably intended to pressure the Court into backing down. At minimum, it was probably intended to encourage some justices to vote against upholding the prosecutor's request. If the vote was split, the President could argue that it was not sufficiently definitive for a matter of this magnitude. Jaworski believed that in the event of a split vote, the President would "go on television and tell the people that the presidency should not be impaired by a divided Court."

We model this situation as a two-player game. Player 1 is Justices Burger and Blackmun, whom we assume will vote together; we therefore treat them as one player. Player 2 is President Nixon.

First, Justices Burger and Blackmun decide how to vote. If they vote to uphold the prosecutor's request, the result is an 8-0 Supreme Court decision in favor of the prosecutor. If they vote to reject the prosecutor's request, the result is a 6-2 Supreme Court decision in favor of the prosecutor.

After the Supreme Court has rendered its decision, President Nixon decides whether to comply by supplying the tapes or to defy the decision. President Nixon's preferences are as follows:

- Best outcome (payoff 4): 6-2 decision, President defies the decision.
- Second-best outcome (payoff 3): 6-2 decision, President supplies the tapes.
- Third-best outcome (payoff 2): 8-0 decision, President supplies the tapes.
- Worst outcome (payoff 1): 8-0 decision, President defies the decision.

Explanation: The President's best outcome is a divided decision that he can defy while claiming the decision is not really definitive. His worst outcome is an 8-0 decision that he then defies; this would probably result in immediate impeachment. As for the two intermediate outcomes, the President is better off with the weaker vote, which should give him some wiggle room.

Justices Burger and Blackmun's preferences are as follows:

- Best outcome (payoff 4): 6-2 decision, President supplies the tapes.
- Second-best outcome (payoff 3): 8-0 decision, President supplies the tapes.
- Third-best outcome (payoff 2): 8-0 decision, President defies the decision.
- Worst outcome (payoff 1): 6-2 decision, President defies the decision.

Explanation: In their best outcome, Burger and Blackmun get to vote their honest legal opinion that the prosecutor's direct appeal to the Court was wrong, but a Constitutional crisis is averted, because the President complies anyway. In their second-best outcome, they vote dishonestly, but they succeed in averting a major Constitutional crisis. In their third-best outcome, the crisis occurs, but because of the strong 8-0 vote, it will probably quickly end. In the worst outcome, the crisis occurs, and because of the weak vote, it may drag out. In addition, in the last outcome, the President may succeed in establishing the principle that a 6-2 Court decision need not be followed, which no member of the Court wants.

(1) Draw a game tree and use backward induction to predict what happened.
(2) Can you think of a plausible way that President Nixon might have gotten a better outcome?

(What actually happened: the Court ruled 8-0 in favor of the prosecutor on July 24. On July 30, President Nixon surrendered the tapes. In early August, a previously unknown tape recorded a few days after the break-in was released. It documented President Nixon and his aide Robert Haldeman formulating a plan to block investigations by having the CIA claim to the FBI (falsely) that national security was involved. On August 9, President Nixon resigned.)

1.14.5 Three Pirates. Three pirates must divide 100 gold doubloons. The doubloons cannot be cut into pieces. Pirate A is the strongest, followed by Pirate B, followed by Pirate C. Because of ancient pirate tradition, the coins are divided in the following manner. First Pirate A proposes a division of the coins. The three pirates then vote on whether to accept the proposed division. If the proposal gets a majority vote, it is accepted, and the game is over. If the proposal fails to get a majority vote, Pirate A is executed.

It is then Pirate B's turn to propose a division of the coins between the two remaining pirates. The same rules apply, with one exception: if the vote is a tie (which can happen this time, because the number of pirates is now even), Pirate B, being the strongest remaining pirate, gets an additional vote to break the tie.

Use backward induction to figure out what Pirate A should propose. You don't have to draw a game tree if you don't want to, but you do have to explain your thinking so that your instructor can follow it.

1.14.6 Grocery Store and Gas Station 1. In a certain town, there are two stores, a grocery store and a gas station. The grocery store charges p_1 dollars per pound for food, and the gas station charges p_2 dollars per gallon for gas. The grocery store sells q_1 pounds of food per week, and the gas station sells q_2 gallons of gas per week. The quantities q_1 and q_2 are related to the prices

p_1 and p_2 as follows:

$$q_1 = 10 - 2p_1 - p_2,$$
$$q_2 = 10 - p_1 - 2p_2.$$

Thus, if the price of food or gas rises, less of *both* is sold.

Let π_1 be the revenue of the grocery store and π_2 be the revenue of the gas station. Both depend on the two stores' choices of p_1 and p_2:

$$\pi_1(p_1, p_2) = q_1 p_1 = (10 - 2p_1 - p_2)p_1 = 10p_1 - 2p_1^2 - p_1 p_2,$$
$$\pi_2(p_1, p_2) = q_2 p_2 = (10 - p_1 - 2p_2)p_2 = 10p_2 - p_1 p_2 - 2p_2^2.$$

We interpret this as a game with two players, the grocery store (Player 1) and the gas station (Player 2). The payoff to each player is its revenue.

Allow p_1 and p_2 to be any real numbers, even negative numbers or numbers that produce negative values for q_1 and q_2.

Suppose the grocery store chooses its price p_1 first, and then the gas station, knowing p_1, chooses its price p_2.

(1) If the grocery store uses backward induction to choose p_1, what price will it choose?

(2) What will be the corresponding p_2, and what will be the revenue of each store?

Partial answer to help keep you on the right track: $p_1 = 2\frac{1}{7}$.

1.14.7 Grocery Store and Gas Station 2. We now change the problem a little. First we change the formulas for q_1 and q_2 to the following more reasonable formulas, which say that when prices get too high, the quantities sold become 0, not negative numbers:

$$q_1 = \begin{cases} 10 - 2p_1 - p_2 & \text{if } 2p_1 + p_2 < 10, \\ 0 & \text{if } 2p_1 + p_2 \geqslant 10, \end{cases}$$

$$q_2 = \begin{cases} 10 - p_1 - 2p_2 & \text{if } p_1 + 2p_2 < 10, \\ 0 & \text{if } p_1 + 2p_2 \geqslant 10. \end{cases}$$

Next we make some reasonable restrictions on the prices. First we assume

$$p_1 \geqslant 0 \quad \text{and} \quad p_2 \geqslant 0.$$

Now we note that if $p_1 \geqslant 0$ and $p_2 > 5$, then q_2 becomes 0. The gas station wouldn't want this, so we assume

$$p_2 \leqslant 5.$$

Finally, we note that if $p_2 \geqslant 0$ and $p_1 > 5$, then q_1 becomes 0. The grocery store wouldn't want this, so we assume

$$p_1 \leqslant 5.$$

The formulas for π_1 and π_2 are now a little different, because the formulas for q_1 and q_2 have changed. The payoffs are now only defined for $0 \leqslant p_1 \leqslant 5$ and $0 \leqslant p_2 \leqslant 5$, and are given by

$$\pi_1(p_1, p_2) = q_1 p_1 = \begin{cases} 10p_1 - 2p_1^2 - p_1 p_2 & \text{if } 2p_1 + p_2 < 10, \\ 0 & \text{if } 2p_1 + p_2 \geqslant 10, \end{cases}$$

$$\pi_2(p_1, p_2) = q_2 p_2 = \begin{cases} 10p_2 - p_1 p_2 - 2p_2^2 & \text{if } p_1 + 2p_2 < 10, \\ 0 & \text{if } p_1 + 2p_2 \geqslant 10. \end{cases}$$

We still assume that the grocery store chooses its price p_1 first, and then the gas station, knowing p_1, chooses its price p_2.

(1) Explain the formulas above for the payoffs.
(2) For a fixed p_1 between 0 and 5, graph the function $\pi_2(p_1, p_2)$, which is a function of p_2 defined for $0 \leqslant p_2 \leqslant 5$. Answer: should be partly an upside-down parabola and partly a horizontal line. Make sure you give a formula for the point at which the graph changes from one to the other.
(3) By referring to the graph you just drew and using calculus, find the gas station's best-response function $p_2 = b(p_1)$, which should be defined for $0 \leqslant p_1 \leqslant 5$. Answer: $b(p_1) = (10 - p_1)/4$.
(4) You are now ready to find p_1 by backward induction. From your formula for π_1 you should be able to see that

$$\pi_1(p_1, b(p_1)) = \begin{cases} (10 - 2p_1 - b(p_1))p_1 & \text{if } 2p_1 + b(p_1) < 10, \\ 0 & \text{if } 2p_1 + b(p_1) \geqslant 10. \end{cases}$$

Use the formula for $b(p_1)$ from part (3) to show that $2p_1 + b(p_1) < 10$ if $0 \leqslant p_1 < 4\frac{2}{7}$, and $2p_1 + b(p_1) \geqslant 10$ if $4\frac{2}{7} \leqslant p_1 \leqslant 5$.
(5) Graph the function $\pi_1(p_1, b(p_1))$, $0 \leqslant p_1 \leqslant 5$. (Again, it is partly an upside-down parabola and partly a horizontal line.)
(6) By referring to the graph you just drew and using calculus, find where $\pi_1(p_1, b(p_1))$ is maximum.

1.14.8 Tax Rate. Government chooses a tax rate x, $0 < x < 1$. Citizens then choose a level of effort y to devote to making money. The resulting level of economic activity a, in trillions of dollars, is

$$a = 4y - 4xy - y^2.$$

Explanation: The term $4y$ expresses the idea that economic activity should be proportional to effort. There are two correction terms. The term $-4xy$ expresses the idea that when the tax rate is high, much effort goes into avoiding taxes, not into economic activity. The term $-y^2$ expresses the idea that too much effort is counterproductive.

The government's payoff is the taxes it collects, computed as tax rate x times level of economic activity a:

$$\pi_1(x, y) = xa = x(4y - 4xy - y^2).$$

The citizens' payoff is the part of economic activity that they keep after taxes:

$$\pi_2(x, y) = a - xa = (1 - x)a = (1 - x)(4y - 4xy - y^2).$$

We regard this as a two-player game. Player 1 is the government. Player 2 is the citizens. Player 1 chooses the tax rate x, then Player 2 observes x and chooses y. The payoffs are given above.

Use backward induction to find the tax rate x that maximizes the government's payoff.

1.14.9 Continuous Ultimatum Game with Inequality Aversion. Players 1 and 2 must divide some Good Stuff. Player 1 offers Player 2 a fraction y, $0 \leqslant y \leqslant 1$, of the Good Stuff. If Player 2 accepts the offer, she gets the fraction y of Good Stuff, and Player 1 gets the remaining fraction $x = 1 - y$. If Player 2 rejects the offer, both players get nothing.

In this game, Player 1 has an interval of possible strategies. We can describe this interval as $0 \leqslant x \leqslant 1$, where x is the fraction Player 1 keeps, or as $0 \leqslant y \leqslant 1$, where y is the fraction Player 1 offers to Player 2. In contrast, Player 2's strategy is a plan, for each y that she might be offered, whether to accept or reject.

The players' payoffs are partly objective (their fraction of the Good Stuff) and partly subjective. Both players are *inequality averse*: they don't like an unequal division. However, each player feels that an inequality favoring the other player is worse than inequality favoring herself. Their payoffs are as follows:

$$u_1(x, y) = x - \begin{cases} \alpha_1(y - x) & \text{if } y \geqslant x, \\ \beta_1(x - y) & \text{if } y < x, \end{cases}$$

$$u_2(x, y) = y - \begin{cases} \alpha_2(x - y) & \text{if } x \geqslant y, \\ \beta_2(y - x) & \text{if } x < y, \end{cases}$$

with $0 < \beta_1 < \alpha_1$ and $0 < \beta_2 < \alpha_2$. Thus Player 1's payoff is the fraction x of Good Stuff that she gets, minus a correction for inequality. The correction is proportional to the difference between the two allocations. If Player 2 gets more ($y > x$), the difference is multiplied by the bigger number α_1; if Player 2 gets less ($y < x$), the difference is multiplied by the smaller number β_1. Player 2's payoff function is similar.

If Player 1's offer is rejected, both players receive 0 payoff.

We assume that Player 2 will accept offers that make her payoff positive *or zero*. Player 1 wants to make an offer that (a) Player 2 will accept, and (b) that leaves Player 1 with the largest possible payoff.

Since $y = 1 - x$ and $x = 1 - y$, we simplify the payoff functions as follows:

$$u_1(x) = x - \begin{cases} \alpha_1(1 - 2x) & \text{if } x \leqslant \frac{1}{2}, \\ \beta_1(2x - 1) & \text{if } x > \frac{1}{2}, \end{cases}$$

$$u_2(y) = y - \begin{cases} \alpha_2(1 - 2y) & \text{if } y \leqslant \frac{1}{2}, \\ \beta_2(2y - 1) & \text{if } y > \frac{1}{2}. \end{cases}$$

(1) Graph Player 1's payoff function on the interval $0 \leqslant x \leqslant 1$ assuming $\beta_1 > \frac{1}{2}$. The graph consists of two line segments that meet at the point $(\frac{1}{2}, \frac{1}{2})$. Your graph should clearly indicate the points with x-coordinates 0 and 1.

(2) Explain briefly: If $\beta_1 > \frac{1}{2}$, then Player 1 offers Player 2 half of the Good Stuff, and Player 2 accepts.

(3) Graph Player 1's payoff function on the interval $0 \leqslant x \leqslant 1$ assuming $\beta_1 < \frac{1}{2}$.

(4) Graph Player 2's payoff function on the interval $0 \leqslant y \leqslant 1$ assuming $\beta_2 < 1$. Use your graph to explain the following statement: When $\beta_2 \leqslant 1$, Player 2 will accept any offer $y \geqslant y^*$ with $y^* = \alpha_2/(1 + 2\alpha_2)$.

(5) If $\beta_1 < \frac{1}{2}$ and $\beta_2 \leqslant 1$, what fraction of the Good Stuff should Player 1 offer to Player 2?

(6) If $\beta_1 < \frac{1}{2}$ and $\beta_2 > 1$, what fraction of the Good Stuff should Player 1 offer to Player 2?

Chapter 2

Eliminating dominated strategies

In this chapter we introduce games in which the players make their choices simultaneously. The well-known Prisoner's Dilemma, which models situations in which cooperation is especially difficult, is a game of this type. We discuss the simplest idea for dealing with these games, which is not to use a strategy when another is available that never gives a worse result and sometimes gives a better one. Using this idea, one can understand why the second-price auction, which is essentially how items are sold on e-Bay, works so well. In the Prisoner's Dilemma, however, this idea leads to an uncooperative, suboptimal outcome.

2.1 Prisoner's Dilemma

Two corporate executives are accused of preparing false financial statements. The prosecutor has enough evidence to send both to jail for one year. However, if one confesses and tells the prosecutor what he knows, the prosecutor will be able to send the other to jail for ten years. In exchange for the help, the prosecutor will let the executive who confesses go free. If both confess, both will go to jail for six years. The executives are held in separate cells and cannot communicate. Each must decide individually whether to talk or refuse.

Since each executive decides what to do without knowing what the other has decided, it is not natural or helpful to draw a game tree. Nevertheless, we can still identify the key elements of a game: players, strategies, and payoffs.

The players are the two executives. Each has the same two strategies: talk or refuse. The payoff to each executive is number of years in jail (preceded by a minus sign, since we want higher payoffs to be more desirable). The payoff to each executive depends on the strategy choices of both executives.

In this two-player game, we can indicate how the strategies determine the payoffs by a matrix:

| | | Executive 2 | |
		refuse	talk
Executive 1	refuse	$(-1, -1)$	$(-10, 0)$
	talk	$(0, -10)$	$(-6, -6)$

The rows of the matrix represent Executive 1's strategies. The columns represent Executive 2's strategies. Each entry of the matrix is an ordered pair of numbers that gives the payoffs to the two players if the corresponding strategies are used. Executive 1's payoff is given first.

Notice the following:

(1) If Executive 2 refuses to talk, Executive 1 gets a better payoff by talking than by refusing. (Look at the two ordered pairs in the first column of the matrix, and compare their first entries: 0 is better than -1.)

(2) If Executive 2 talks, Executive 1 still gets a better payoff by talking than by refusing. (Look at the two ordered pairs in the second column of the matrix, and compare their first entries: -6 is better than -10.)

Thus, no matter what Executive 2 does, Executive 1 gets a better payoff by talking than by refusing. Executive 1's strategy of talking *strictly dominates* his strategy of refusing: it gives a better payoff to Executive 1 no matter what Executive 2 does.

Of course, Executive 2's situation is identical: his strategy of talking gives him a better payoff no matter what Executive 1 does. (In each row of the matrix, compare the second entries in the two ordered pairs. The one in the second column is better.)

Thus we expect both executives to talk. Unfortunately for them, the result is that they both go to jail for 6 years. Had they both refused to talk, they would have gone to jail for only one year.

Prosecutors like playing this game. Defendants don't like it much. Hence there have been attempts over the years by defendants' attorneys and friends to change the game.

For example, if the Mafia were involved, it might have told the two executives in advance: "If you talk, something bad could happen to your child." Suppose each executive believes this warning and considers something bad happening to his child to be equivalent to six years in prison. The payoffs in the game are changed as follows:

| | | Executive 2 | |
		refuse	talk
Executive 1	refuse	$(-1, -1)$	$(-10, -6)$
	talk	$(-6, -10)$	$(-12, -12)$

Now, for both executives, the strategy of refusing to talk strictly dominates the strategy of talking. Thus we expect both executives to refuse to talk, so both go to jail for only one year. The Mafia's threat sounds cruel. In this instance, however, it helped the two executives achieve a better outcome for themselves than they could achieve on their own.

Prosecutors don't like the second version of the game. One mechanism they have of returning to the first version is to offer "witness protection" to prisoners who talk. In a witness protection program, the witness and his family are given new identities in a new town. If the prisoner believes that the Mafia is thereby prevented from carrying out its threat, the payoffs return to something close to those of the original game.

Another way to change the game of Prisoner's Dilemma is by additional rewards. For example, the Mafia might work hard to create a culture in which prisoners who don't talk are honored by their friends, and their families are taken care of. If the two executives buy into this system and consider the rewards of not talking to be worth five years in prison, the payoffs become the following:

		Executive 2	
		refuse	talk
Executive 1	refuse	$(4,4)$	$(-5,0)$
	talk	$(0,-5)$	$(-6,-6)$

Once again refusing to talk strictly dominates talking.

The punishments or rewards that can change the dominant strategy in the Prisoner's Dilemma from talk to refuse can be completely internal. A prisoner may simply feel badly about himself if he selfishly betrays his fellow prisoner, or may feel good about himself if he doesn't. If these feelings are important, we have to take them into account when we assign his payoffs.

The Prisoner's Dilemma is the best-known and most studied model in game theory. It models many common situations in which cooperation is advantageous but difficult to achieve. We illustrate this point in Sections 2.4 and 2.5, where we discuss the Israeli-Palestinian conflict and the problem of global warming. You may also want to look at the Wikipedia page on the Prisoner's Dilemma (http://en.wikipedia.org/wiki/Prisoners_dilemma).

2.2 Games in normal form

A game in normal form consists of:

(1) A finite set P of players. We will usually take $P = \{1, \ldots, n\}$.
(2) For each player i, a set S_i of available strategies.

Let $S = S_1 \times \cdots \times S_n$. An element of S is an n-tuple $(s_1, \ldots, s_n)$, where each s_i is a strategy chosen from the set S_i. Such an n-tuple $(s_1, \ldots, s_n)$ is called a *strategy profile*. It represents a choice of strategy by each of the n players.

(3) For each player i, a *payoff function* $\pi_i : S \to \mathbb{R}$.

In the Prisoner's Dilemma,

$$P = \{1, 2\}, \quad S_1 = \{\text{refuse,talk}\}, \quad S_2 = \{\text{refuse,talk}\},$$

and S is a set of four ordered pairs, namely, (refuse,refuse), (refuse,talk), (talk,refuse), and (talk,talk). As to the payoff functions, we have, for example, $\pi_1(\text{refuse,talk}) = -10$ and $\pi_2(\text{refuse,talk}) = 0$.

If there are two players, Player 1 has m strategies, and Player 2 has n strategies, then a game in normal form can be represented by an $m \times n$, matrix as in the previous section. Each row of the matrix corresponds to a strategy of Player 1, and each column to a strategy of Player 2. Each entry in the matrix corresponds to a strategy profile (i.e., a choice of strategy by each player). The entry is the ordered pair of payoffs associated to the strategy profile.

2.3 Dominated strategies

For a game in normal form, let s_i and s_i' be two of Player i's strategies.

- We say that s_i *strictly dominates* s_i' if, for every choice of strategies by the other players, the payoff to Player i from using s_i is greater than the payoff to Player i from using s_i'.
- We say that s_i *weakly dominates* s_i' if, for every choice of strategies by the other players, the payoff to Player i from using s_i is at least as great as the payoff to Player i from using s_i'; and, for some choice of strategies by the other players, the payoff to Player i from using s_i is greater than the payoff to Player i from using s_i'.

To illustrate these definitions, consider a two-player game in normal form in which Player 1 has two strategies a and b; Player 2 has four strategies c, d, e, and f; and the payoff matrix is the following:

		Player 2			
		c	*d*	*e*	*f*
Player 1	*a*	$(1, 2)$	$(-1, 2)$	$(2, 4)$	$(-1, 3)$
	b	$(4, 1)$	$(1, 1)$	$(3, 1)$	$(0, 2)$

Player 1's strategy b strictly dominates her strategy a, because it gives her a better payoff no matter what strategy Player 2 uses: $4 > 1$, $1 > -1$, $3 > 2$, and $0 > -1$. Similarly, Player 2's strategy f strictly dominates both c and d, because $3 > 2$ and $2 > 1$. Player 2's strategy e weakly dominates both c and d: in both cases, it gives Player 2 a better payoff when Player 1 uses a, and the same payoff when Player 1 uses b. However, Player 2's strategy d does not weakly dominate c, since there is no case in which it gives Player 2 a better payoff. Similarly, c does not weakly dominate d. Player 2's strategy f does not weakly dominate e, since there is a case in which f gives a worse payoff than e (when Player 1 uses a). Similarly, Player 2's strategy e does not weakly dominate f.

If a strategy s_i strictly dominates another strategy s_i', then s_i also weakly dominates s_i'. If s_i does not weakly dominate s_i', then s_i also does not strictly dominate s_i'.

As mentioned in Section 1.4, game theorists often assume that players are rational. For a game in normal form, rationality is often taken to imply the following:

- Suppose one of Player i's strategies s_i weakly dominates another of her strategies s_i'. Then Player i will not use the strategy s_i'.

This is the assumption we used to analyze the Prisoner's Dilemma. Actually, in that case, we only needed to eliminate strictly dominated strategies.

A prisoner's dilemma occurs when (i) each player has a strategy that strictly dominates all her other strategies, but (ii) each player has another strategy such that, if all players were to use this alternative, all players would receive higher payoffs than those they receive when they all use their dominant strategies.

2.4 Israelis and Palestinians

Henry Kissinger was National Security Advisor and later Secretary of State during the administrations of Richard Nixon and Gerald Ford. Previously he was a professor of international relations at Harvard. In his view, the most important contribution of the game theory point of view in international relations was that it forced you to make an explicit model of the situation you wanted to understand.

Let's look at the Israeli-Palestinian conflict with this opinion of Kissinger's in mind. In a war between Israel and the neighboring Arab countries in 1967, the Israeli army occupied both Gaza and the West Bank, as well as other territories. The West Bank especially was seen by many Israelis as being a natural part of the state of Israel for both religious reasons (the Jewish heartland in

Biblical times was in what is now the West Bank) and military reasons. Considerable Jewish settlement took place in the West Bank, and to a lesser extent in Gaza, after 1968, with the goal of retaining at least part of these territories in an eventual resolution of the conflict.

In 2000, negotiations between Israeli Prime Minister Ehud Barak and Palestinian leader Yasser Arafat, with the mediation of U.S. President Bill Clinton, perhaps came close to resolving the conflict. Barak offered to remove most of the Israeli settlements and allow establishment of a Palestinian state. Arafat rejected the offer. The level of conflict between the two sides increased greatly. In 2005, the Israelis abandoned their settlements in Gaza and ended their occupation of that region.

During the first decade of the twenty-first century, discussion of this conflict often focused on two issues: control of the West Bank and terrorism. Most proposals for resolving the conflict envisioned a trade-off in which the Israelis would end their occupation of the West Bank and the Palestinians would stop terrorism, the means by which they carried on their conflict with Israel.

In a simple model of the conflict, the Israelis had two possible strategies: end the occupation or continue to occupy the West Bank. The Palestinians also had two possible strategies: end terrorism or continue terrorism. What were the payoffs?

The Israelis certainly valued both keeping the West Bank and an end to Palestinian terrorism. The Palestinians certainly valued ending the Israeli occupation of the West Bank. We will assume that the Palestinians also valued retaining their freedom to continue terrorism. The reason is that for the Palestinians, giving up terrorism essentially meant giving up hope of regaining the pre-1967 territory of Israel, which was the home of many Palestinians, and which many Palestinians feel is rightfully their territory.

Let's consider two ways to assign payoffs that are consistent with this assumption.

1. At the time of the negotiations between Barak and Arafat, Barak apparently considered an end to terrorism to be of greater value than continued occupation of the West Bank, since he was willing to give up the latter in exchange for the former. Therefore we will assign the Israelis 2 points if terrorism ends, and −1 point if they end the occupation of the West Bank.

Arafat apparently considered retaining the freedom to engage in terrorism to be of greater value than ending the Israeli occupation of the West Bank, since he was not willing to give up the former to achieve the latter. Therefore, we assign the Palestinians −2 points if they end terrorism, and 1 point if the Israelis end their occupation of the West Bank. We get the following game in normal form:

| | | Palestinians ||
		end terrorism	continue terrorism
Israelis	end occupation	$(1, -1)$	$(-1, 1)$
	continue occupation	$(2, -2)$	$(0, 0)$

The payoff matrix shows that for the Israelis, continuing the occupation strictly dominates ending it, and for the Palestinians, continuing terrorism strictly dominates ending it. These strategies yield the actual outcome of the negotiations.

This game is not a prisoner's dilemma. In a prisoner's dilemma, each player has a dominant strategy, but the use of the dominated strategies by each player would result in a higher payoff to both. Here, if each player uses his dominated strategy, the Israeli outcome improves, but the Palestinian outcome is worse.

2. The previous assignment of payoffs was appropriate for Israeli "moderates" and Palestinian "radicals." We will now assign payoffs on the assumption that both sides are "moderate." The Israeli payoffs are unchanged. The Palestinians are now assumed to value ending the occupation of the West Bank above keeping the freedom to engage in terrorism. We therefore assign the Palestinians -1 point if they end terrorism, and 2 points if the Israelis end their occupation of the West Bank. We get the following game in normal form:

| | | Palestinians ||
		end terrorism	continue terrorism
Israelis	end occupation	$(1, 1)$	$(-1, 2)$
	continue occupation	$(2, -1)$	$(0, 0)$

For the Israelis, continuing the occupation still strictly dominates ending it, and for the Palestinians, continuing terrorism still strictly dominates ending it. This indicates that even "moderate" governments on both sides would have difficulty resolving the conflict.

The game is now a prisoner's dilemma: if both sides use their dominated strategies, there will be a better outcome for both, namely, an end to both the Israeli occupation of the West Bank and to Palestinian terrorism. As in the original Prisoner's Dilemma, this outcome is not easy to achieve. Also as in the original Prisoner's Dilemma, one solution is for an outside player to change the payoffs by supplying punishments or rewards, as the Mafia could there. In the context of the Israeli-Palestinian conflict, the most plausible such outside player is the United States.

Of course, once one considers whether the United States wants to become involved, one has a three-player game. If one considers subgroups within the Israelis and Palestinians (for example, Israeli extremists, or the radical Palestinian group Hamas), the game becomes even more complicated.

The Israeli-Palestinian situation illustrates the dilemma of cooperation. Both the Israelis and the Palestinians are asked to help the other at a cost to themselves. To generalize this situation, let us suppose that Player 1 can confer a benefit $b > 0$ on Player 2 at a cost of $a > 0$ to herself. Similarly, Player 2 can confer a benefit $d > 0$ on Player 1 at a cost of $c > 0$ to herself. We get the following game in normal form:

		Player 2	
		help	don't help
Player 1	help	$(d - a, b - c)$	$(-a, b)$
	don't help	$(d, -c)$	$(0, 0)$

For both players, don't help strictly dominates help. However, if $d > a$ and $b > c$ (i.e., if for both players, the benefit from getting help is greater than the cost of helping), then we have a prisoner's dilemma: if both players help, both will be better off.

2.5 Global Warming

Ten countries are considering fighting global warming. Each country must choose to spend an amount x_i to reduce its carbon emissions, where $0 \leqslant x_i \leqslant 1$. The total benefits produced by these expenditures equal twice the total expenditures: $2(x_1 + \cdots + x_{10})$. Each country receives $\frac{1}{10}$ of the benefits.

This game has ten players, the ten countries. The set of strategies available to country i is just the closed interval $0 \leqslant x_i \leqslant 1$. A strategy profile is therefore a 10-tuple $(x_1, \ldots, x_{10})$, where $0 \leqslant x_i \leqslant 1$ for each i. The ith country's payoff function is its benefits minus its expenditures:

$$\pi_i(x_1, \ldots, x_{10}) = \frac{1}{10} \cdot 2(x_1 + \cdots + x_{10}) - x_i = \frac{1}{5}(x_1 + \cdots + x_{10}) - x_i.$$

We will show that for each country, the strategy $x_i = 0$ (spend nothing to fight global warming) strictly dominates all its other strategies.

We just show this for country 1, since the argument for any other country is the same. Let $x_1 > 0$ be a different strategy for country 1. Let $x_2, \ldots, x_n$ be any strategies for the other countries. Then

$$\pi_1(0, x_2, \ldots, x_{10}) - \pi_1(x_1, x_2, \ldots, x_{10})$$
$$= \left(\frac{1}{5}(0 + x_2 + \cdots + x_{10}) - 0\right) - \left(\frac{1}{5}(x_1 + x_2 + \cdots + x_{10}) - x_1\right)$$
$$= -\frac{1}{5}x_1 + x_1 = \frac{4}{5}x_1 > 0.$$

Thus we expect each country to spend nothing to fight global warming, and each country to get a payoff of 0.

If all countries could somehow agree to spend 1 each to fight global warming, each country's payoff would be $\frac{1}{5}(1 + \cdots + 1) - 1 = 2 - 1 = 1$, and each country would be better off. In fact, each country would receive benefits of 2 in return for expenditures of 1, an excellent deal.

Nevertheless, each country would be constantly tempted to cheat. A reduction in country i's expenditures by y_i dollars reduces total benefits to all countries by $2y_i$ dollars, but only reduces benefits to country i by $\frac{1}{5}y_i$ dollars.

This example suggests that the problem of global warming is a type of prisoner's dilemma.

Of course, one can try to change the game by changing the payoffs with punishments or rewards. For example, one might try to raise the environmental consciousness of people around the world by a publicity campaign. Then perhaps governments that fight global warming would get the approval of their own people and the approval of others around the world, which they might see as a reward. In addition, governmental leaders might get subjective rewards by doing what they feel is the right thing.

Games such as the one described in this section are called *public goods games*. In a public goods game, when a player cooperates, she adds more to the total payoffs of all players than her cost of cooperating, but her cost of cooperating is greater than her individual share of the payoffs. Public goods games are one type of *social dilemma*. In a social dilemma, all players gain when all cooperate, but each has an incentive to defect, which will give her a gain at the expense of the others.

2.6 Hagar's Battles

There are ten villages with values $a_1 < a_2 < \cdots < a_{10}$. There are two players. Player 1 has n_1 soldiers, and Player 2 has n_2 soldiers, with $0 < n_1 < 10$ and $0 < n_2 < 10$. Each player independently decides which villages to send his soldiers to. A player is not allowed to send more than one soldier to a village.

A player wins a village if he sends a soldier there but his opponent does not.

A player's score is the sum of the values of the villages he wins, minus the sum of the values of the villages his opponent wins. Each player wants to maximize his score (not just beat his opponent).

Where should you send your soldiers?

Since each player decides where to send his soldiers without knowledge of the other player's decision, we model this game as a game in normal form. To do that, we must describe precisely the players, the strategies, and the payoff functions.

- Players. There are two.
- Strategies. The villages are numbered from 1 to 10. A strategy for Player i is just a set of n_i numbers between 1 and 10. The numbers represent the n_i different villages to which he sends his soldiers. Thus if S_i is the set of all of Player i's strategies, an element s_i of S_i is simply a set of n_i numbers between 1 and 10.
- Payoff functions. A player's payoff in this game is his score as previously defined.

A neat way to analyze this game is to find a nice formula for the payoff function. Let's look at an example. Suppose $n_1 = n_2 = 3$, $s_1 = \{6, 8, 10\}$, and $s_2 = \{7, 9, 10\}$. Player 1 wins villages 6 and 8, and Player 2 wins villages 7 and 9. Thus Player 1's payoff is $(a_6 + a_8) - (a_7 + a_9)$, and Player 2's payoff is $(a_7 + a_9) - (a_6 + a_8)$. Since $a_6 < a_7$ and $a_8 < a_9$, Player 2's score is higher.

We could also calculate Player i's payoff by adding the values of all the villages to which he sends his soldiers and subtracting the values of all the villages to which his opponent sends his soldiers. Then we would have

- Player 1's payoff $= (a_6 + a_8 + a_{10}) - (a_7 + a_9 + a_{10}) = (a_6 + a_8) - (a_7 + a_9)$.
- Player 2's payoff $= (a_7 + a_9 + a_{10}) - (a_6 + a_8 + a_{10}) = (a_7 + a_9) - (a_6 + a_8)$.

Clearly this method of calculating payoffs always works. Thus we have the following formulas for the payoff functions:

$$\pi_1(s_1, s_2) = \sum_{j \in s_1} a_j - \sum_{j \in s_2} a_j,$$
$$\pi_2(s_1, s_2) = \sum_{j \in s_2} a_j - \sum_{j \in s_1} a_j.$$

We claim that for each player, the strategy of sending his n_i soldiers to the n_i villages of highest values *strictly dominates* all his other strategies.

We will just show that for Player 1, the strategy of sending his n_1 soldiers to the n_1 villages of highest values strictly dominates all his other strategies. The argument for Player 2 is the same.

Let s_1 be the set of the n_1 highest numbers between 1 and 10. (For example, if $n_1 = 3$, $s_1 = \{8, 9, 10\}$). Let s_1' be a different strategy for Player 1 (i.e., a different set of n_1 numbers between 1 and 10). Let s_2 be any strategy for Player 2 (i.e., any set of n_2 numbers between 1 and 10). We must show that

$$\pi_1(s_1, s_2) > \pi_1(s_1', s_2),$$

or, equivalently, that $\pi_1(s_1, s_2) - \pi_1(s_1', s_2) > 0$.

We have

$$\pi_1(s_1, s_2) = \sum_{j \in s_1} a_j - \sum_{j \in s_2} a_j,$$

$$\pi_1(s_1', s_2) = \sum_{j \in s_1'} a_j - \sum_{j \in s_2} a_j.$$

Therefore

$$\pi_1(s_1, s_2) - \pi_1(s_1', s_2) = \sum_{j \in s_1} a_j - \sum_{j \in s_1'} a_j.$$

This is clearly positive: the sum of the n_1 biggest numbers between 1 and 10 is greater than the sum of some other n_1 numbers between 1 and 10.

2.7 Second-price auctions

An item is to be sold at auction. Each bidder submits a sealed bid. All the bids are opened. The object is sold to the highest bidder, but the price is the bid of the second-highest bidder. (If two or more bidders submit equal highest bids, that is the price, and one of those bidders is chosen by chance to buy the object. However, we ignore this possibility in our analysis.)

If you are a bidder at such an auction, how much should you bid?

Clearly the outcome of the auction depends not only on what you do, but also on what the other bidders do. Thus we can think of the auction as a game. Since the bidders bid independently, without knowledge of the other bids, we try to model this auction as a game in normal form. We must describe precisely the players, the strategies, and the payoff functions.

- Players. Suppose there are n bidders.
- Strategies. The ith player's strategy is simply her bid, which we denote b_i. At this point we must decide whether to allow just integer bids, arbitrary real number bids, or something else. Let's try allowing the bids b_i to be any nonnegative real number. The ith player's set of available bids is then $S_i = [0, \infty)$. As in the Global Warming game of Section 2.5, each set S_i is an interval.
- Payoff functions. A reasonable idea is that the payoff to player i is 0 unless she wins the auction, in which case the payoff to Player i is the value of the object to Player i minus the price she has to pay for it. Thus the payoff to Player i depends on
  - the value of the object to Player i, which we denote v_i;
  - the bid of Player i, b_i; and
  - the highest bid of the other players, which we denote $h_i = \max\{b_j : j \neq i\}$.

The formula is

$$\pi_i(b_1,\ldots,b_n) = \begin{cases} 0 & \text{if } b_i < h_i, \\ v_i - h_i & \text{if } h_i < b_i. \end{cases}$$

(Recall that we are ignoring the possibility that two bidders submit equal highest bids; i.e., we ignore the possibility that $b_i = h_i$.)

We claim that for Player i, the strategy v_i weakly dominates every other strategy. In other words, you should bid exactly what the object is worth to you. (This is the great thing about second-price auctions.)

To show this, we just show that for Player 1, the strategy v_1 weakly dominates every other strategy. The argument for any other player is the same.

Let $b_1 \neq v_1$ be another possible bid by Player 1. We must show two things:

(1) If $b_2,\ldots,b_n$ are any bids by the other players, then

$$\pi_1(v_1,b_2,\ldots,b_n) \geq \pi_1(b_1,b_2,\ldots,b_n).$$

(2) There are some bids $b_2,\ldots,b_n$ by the other players such that

$$\pi_1(v_1,b_2,\ldots,b_n) > \pi_1(b_1,b_2,\ldots,b_n).$$

To show (1) and (2), let $h_1 = \max(b_2,\ldots,b_n)$. As already mentioned, we do not want to consider the possibility that the top two bids are equal, so we assume $v_1 \neq h_1$ and $b_1 \neq h_1$. When Player 1 bids v_1, the outcome of the auction depends on whether $v_1 < h_1$ or $h_1 < v_1$; similarly, when Player 1 bids b_1, the outcome of the auction depends on whether $b_1 < h_1$ or $h_1 < b_1$. This gives four possibilities to look at. We show them in a table:

Relation of v_1 to h_1	Relation of b_1 to h_1	$\pi_1(v_1,b_2,\ldots,b_n)$	$\pi_1(b_1,b_2,\ldots,b_n)$
$v_1 < h_1$	$b_1 < h_1$	0	0
$v_1 < h_1$	$h_1 < b_1$	0	$v_1 - h_1 < 0$
$h_1 < v_1$	$b_1 < h_1$	$v_1 - h_1 > 0$	0
$h_1 < v_1$	$h_1 < b_1$	$v_1 - h_1 > 0$	$v_1 - h_1 > 0$

In every case, $\pi_1(v_1,b_2,\ldots,b_n) \geq \pi_1(b_1,b_2,\ldots,b_n)$. This shows (1).

To show (2) we must consider separately the cases $b_1 < v_1$ and $b_1 > v_1$.

If $b_1 < v_1$, the third line of the table shows that whenever $b_1 < h_1 < v_1$, $\pi_1(v_1,b_2,\ldots,b_n) > \pi_1(b_1,b_2,\ldots,b_n)$. In words, you should not make a bid b_1 below your value v_1 for the object, because if the highest of the other bids is between b_1 and v_1, bidding v_1 will win the auction and get a positive payoff, whereas bidding b_1 would lose the auction and get a zero payoff.

On the other hand, if $b_1 > v_1$, the second line of the table shows that whenever $v_1 < h_1 < b_1$, $\pi_1(v_1, b_2, \ldots, b_n) > \pi_1(b_1, b_2, \ldots, b_n)$. In words, you should not make a bid b_1 above your value v_1 for the object, because if the highest of the other bids is between v_1 and b_1, bidding v_1 will lose the auction and get a zero payoff, whereas bidding b_1 would win the auction and get a negative payoff.

There is a Wikipedia page about second-price auctions: http://en.wikipedia .org/wiki/Sealed_second-price_auction.

2.8 Iterated elimination of dominated strategies

With games in extensive form, the assumption of rationality led to the idea of not choosing a move if one that yielded a higher payoff was available. This notion inspired the idea of repeatedly eliminating such moves, thereby repeatedly simplifying the game, a procedure we called backward induction.

With games in normal form, the assumption of rationality leads to the idea of not using a dominated strategy. If we remove a dominated strategy from a game in normal form, we obtain a game in normal form with one less strategy. If the smaller game has a dominated strategy, it can then be removed. This procedure, known as *iterated elimination of dominated strategies*, can be repeated until no dominated strategies remain. The result is a smaller game to analyze.

If the smaller game includes only one strategy s_i^* for Player i, s_i^* is called a *dominant strategy* for Player i. If the smaller game includes only one strategy s_i^* for *every* player, the strategy profile $(s_1^*, \ldots, s_n^*)$ is called a *dominant strategy equilibrium*.

Iterated elimination of strictly dominated strategies produces the same reduced game in whatever order it is done. However, we shall see that iterated elimination of weakly dominated strategies can produce different reduced games when done in different orders.

2.9 The Battle of the Bismarck Sea

The following description of the Battle of the Bismarck Sea is drastically simplified. For a fuller story, see the Wikipedia page (http://en.wikipedia.org/ wiki/Battle_of_the_Bismarck_Sea).

In 1943, during the Second World War, the Japanese General Imamura wanted to reinforce a base on the island of New Guinea. The supply convoy could take either a rainy northern route or a sunny southern route. The U.S. General Kenney knew the day the convoy would sail and wanted to bomb it. He only had enough reconnaissance aircraft to search one route per day. The northern route was too rainy for bombing one day in three, although it could still be searched by the reconnaissance aircraft. Sailing time was three days.

General Imamura, who was aware that Kenney knew when the convoy would sail, had to decide which route to take. General Kenney had to decide which route to search on that day.

The payoff to General Kenney is the number of days his forces are able to bomb the convoy. The payoff to General Imamura is minus this number. The payoff matrix is

		Imamura sail north	sail south
Kenney	search north	$(1\frac{1}{3}, -1\frac{1}{3})$	$(1\frac{1}{2}, -1\frac{1}{2})$
	search south	$(1, -1)$	$(2, -2)$

The matrix can be explained as follows:

- If the Americans search the correct route, they will on average spend a day finding the Japanese and will have two days to bomb. However, if the route is the northern one, it will rain one-third of the time, leaving $1\frac{1}{3}$ days to bomb.
- If the Americans search the wrong route, at the end of the first day, they will not have found the Japanese and will know they searched the wrong route. On day 2 they will search the other route, and on average find the Japanese after $\frac{1}{2}$ day. (There is less territory to search, since the Japanese have made a day's progress at the end of the first day.) The Americans will have $1\frac{1}{2}$ days to bomb. However, if the Japanese took the northern route, it will rain one-third of the time, leaving one day to bomb.

Neither of General Kenney's strategies dominates the other. For General Imamura, however, sailing north strictly dominates sailing south. We therefore eliminate Imamura's strategy sail south. The resulting game has two strategies for Kenney but only one for Imamura. In this smaller game, Kenney's strategy search north strictly dominates search south. We therefore eliminate search south. What remains is sail north for Imamura, and search north for Kenney. This is in fact what happened.

2.10 Normal form of a game in extensive form with complete information

Recall that for a game in extensive form, a player's strategy is a plan for what action to take in every situation that the player might encounter. We can convert a game in extensive form to one in normal form by simply listing the possible strategies for each of the n players and then associating to each strategy profile the resulting payoffs.

For example, consider the game of Big Monkey and Little Monkey described in Section 1.5. Big Monkey has two strategies, wait (w) and climb (c). Little Monkey has four strategies:

- ww: if Big Monkey waits, wait; if Big monkey climbs, wait.
- wc: if Big Monkey waits, wait; if Big monkey climbs, climb.
- cw: if Big Monkey waits, climb; if Big monkey climbs, wait.
- cc: if Big Monkey waits, climb; if Big monkey climbs, climb.

The normal form of this game has the following payoff matrix:

		Little Monkey			
		ww	wc	cw	cc
Big Monkey	w	$(0,0)$	$(0,0)$	$(9,1)$	$(9,1)$
	c	$(4,4)$	$(5,3)$	$(4,4)$	$(5,3)$

2.11 Big Monkey and Little Monkey 2

The matrix in the previous section is bigger than most of those we have looked at previously (2×4 instead of 2×2). To find dominated strategies for Player 1 in such a game, it is helpful to remember:

(1) Player 1's strategy s_1 strictly dominates his strategy s_1' if and only if, when you compare the ordered pairs in the s_1 row to those in the s_1' row, the first entry of each pair in the s_1 row is greater than the first entry of the corresponding pair in the s_1' row (i.e., the pair in the same column).

(2) Player 1's strategy s_1 weakly dominates his strategy s_1' if and only if, when you compare the ordered pairs in the s_1 row to those in the s_1' row, the first entry of each pair in the s_1 row is greater than or equal to the first entry of the corresponding pair in the s_1' row (i.e., the pair in the same column); and for at least one ordered pair in the s_1 row, the first entry is greater than the first entry of the corresponding pair in the s_1' row.

In our matrix, neither of Player 1's strategies strictly or weakly dominates the other: $4 > 0$ and $5 > 0$, but $9 > 4$ and $9 > 5$.

To find dominated strategies for Player 2:

(1) Player 2's strategy s_2 strictly dominates his strategy s_2' if and only if, when you compare the ordered pairs in the s_2 column to those in the s_2' column, the second entry of each pair in the s_2 column is greater than the second entry of the corresponding pair in the s_2' column (i.e., the pair in the same row).

(2) Player 2's strategy s_2 weakly dominates his strategy s_2' if and only if, when you compare the ordered pairs in the s_2 column to those in the s_2' column, the second entry of each pair in the s_2 column is greater than or equal to the second entry of the corresponding pair in the s_2' column (i.e., the pair in the same row); and for at least one ordered pair in the s_2 column, the second entry is greater than the second entry of the corresponding pair in the s_2' column.

In our matrix, ww weakly dominates wc ($0 = 0$ and $4 > 3$); cw weakly dominates ww ($1 > 0$ and $4 = 4$); cw strictly dominates wc ($1 > 0$ and $4 > 3$); cc weakly dominates wc ($1 > 0$ and $3 = 3$); and cw weakly dominates cc ($1 = 1$ and $4 > 3$). However, when we compare ww and cc, we see that neither strictly or weakly dominates the other.

In the game of Big Monkey and Little Monkey with Big Monkey going first, iterated elimination of weakly dominated strategies produces different reduced games when done in different orders.

Here is one way of doing iterated elimination of weakly dominated strategies in the game:

(1) Eliminate Little Monkey's strategy ww, because it is weakly dominated by cw; and eliminate Little Monkey's strategy wc, because it is weakly dominated by cc.

(2) Eliminate Little Monkey's strategy cc, because it is weakly dominated by cw.

(3) Eliminate Big Monkey's strategy c, because in the reduced 2×1 game it is dominated by w.

(4) What remains is the 1×1 game in which Big Monkey's strategy is w and Little Monkey's strategy is cw. Thus each is a dominant strategy, and (w, cw) is a dominant strategy equilibrium.

These are the strategies we found for Big Monkey and Little Monkey by backward induction.

However, here is another way of doing iterated elimination of weakly dominated strategies in this game:

(1) As before, begin by eliminating Little Monkey's strategy ww, because it is weakly dominated by cw; and then eliminate Little Monkey's strategy wc, because it is weakly dominated by cc.

(2) Eliminate Big Monkey's strategy c, because in the reduced 2×2 game it is dominated by w.

(3) We are left with a 2×1 game in which no more strategies can be eliminated. The remaining strategy profiles are (w, cw) (found before) and (w, cc). This way of doing iterated elimination shows that w is a dominant strategy for Big Monkey, but it does not show that cw is a dominant strategy for Player 2.

2.12 Backward induction and iterated elimination of dominated strategies

For a game in extensive form, each way of going through the backward induction procedure is equivalent to a corresponding way of performing iterated elimination of weakly dominated strategies in the normal form of the same game.

We now show that in the game of Big Monkey and Little Monkey with Big Monkey going first, one way of doing backward induction corresponds to the first way of doing iterated elimination of weakly dominated strategies described in the previous section.

(1) Suppose you begin backward induction by noting that if Big Monkey waits, Little Monkey should climb, and reduce the game tree accordingly. In iterated elimination of weakly dominated strategies, this corresponds to eliminating Little Monkey's strategy ww, because it is weakly dominated by cw (cw gives Little Monkey a better payoff than ww when Big Monkey climbs, and gives the same payoff when Big Monkey waits); and then eliminating Little Monkey's strategy wc, because it is weakly dominated by cc. The payoff matrix of the reduced game has two rows and just two columns (cw and cc).

(2) The second step in backward induction is to note that if Big Monkey climbs, Little Monkey should wait, and reduce the game tree accordingly. In iterated elimination of weakly dominated strategies, this corresponds to eliminating Little Monkey's strategy cc in the 2×2 game, because it is weakly dominated by cw. The payoff matrix of the reduced game has two rows and just the cw column.

(3) The last step in backward induction is to use the reduced game tree to decide that Big Monkey should wait. In iterated elimination of weakly dominated strategies, this corresponds to eliminating Big Monkey's strategy c, because, in the reduced game with only the cw column, it is dominated by w.

We now describe the correspondence between backward induction and iterated elimination of weakly dominated strategies for games in extensive form with two players. The general situation just requires more notation.

Consider a game in extensive form with two players:

- Player 1 moves at nodes c_i, $1 \leqslant i \leqslant p$. At each node c_i she has available a set M_i of moves.
- Player 2 moves at nodes d_j, $1 \leqslant j \leqslant q$. At each node d_j she has available a set N_j of moves.
- A strategy for Player 1 is a choice at each of her nodes of one of the moves available at that node. Thus Player 1's strategy set is $M = M_1 \times \cdots \times M_p$, and a strategy for Player 1 is an ordered p-tuple $(m_1, \ldots, m_p)$ with each $m_i \in M_i$.
- Similarly, Player 2's strategy set is $N = N_1 \times \cdots \times N_q$. A strategy for Player 2 is an ordered q-tuple $(n_1, \ldots, n_q)$ with each $n_j \in N_j$.
- The normal form of the game associates to each pair $(m, n) \in M \times N$ payoffs $\pi_1(m, n)$ and $\pi_2(m, n)$. To determine these payoffs, the game in extensive form is played with the strategies m and n. It ends in a uniquely defined terminal node, whose payoffs are then used.

We consider a backward induction in the extensive form of the game. Assume the players' nodes are numbered so that in the backward induction, each player's nodes are reached in the order last to first.

1. For definiteness, suppose the first node treated in the backward induction is Player 1's node c_p. Each move in M_p ends in a terminal vertex. Let m_p^* be the move in M_p that gives the greatest payoff to Player 1. We assume m_p^* is unique.

Backward induction records the fact that at node c_p, Player 1 will choose m_p^*, deletes from the game tree all moves that start at c_p, and assigns to the now-terminal node c_p the payoffs previously assigned to the end of m_p^*.

The corresponding step in iterated elimination of weakly dominated strategies is to remove all of Player 1's strategies $(m_1, \ldots, m_{p-1}, m_p)$ with $m_p \neq m_p^*$. The reason is that each such strategy is weakly dominated by $(m_1, \ldots, m_{p-1}, m_p^*)$. Against any strategy of Player 2, the latter gives a better payoff if play reaches node c_p, and the same payoff if it does not.

2. Assume now that backward induction has reached Player 1's nodes $c_{k+1}, \ldots, c_p$ and Player 2's nodes $d_{l+1}, \ldots, d_q$. For definiteness, suppose the next node treated is Player 1's node c_k. At this point in the backward induction, each move in M_k ends in a terminal vertex. Let m_k^* be the move in M_k that gives the greatest payoff to Player 1. We assume m_k^* is unique.

Backward induction records the fact that at node c_k, Player 1 will choose m_k^*, deletes from the game tree all moves that start at c_k, and assigns to the now-terminal node c_k the payoffs previously assigned to the end of m_k^*.

At the corresponding step in iterated elimination of weakly dominated strategies, the remaining strategies of Player 1 are those of the form $(m_1, \ldots, m_k, m_{k+1}^*, \ldots, m_p^*)$, and the remaining strategies of Player 2 are those of the form $(n_1, \ldots, n_l, n_{l+1}^*, \ldots, n_q^*)$. We now remove all of Player 1's strategies $(m_1, \ldots, m_{k-1}, m_k, m_{k+1}^*, \ldots, m_p^*)$ with $m_k \neq m_k^*$. The reason is that each such strategy is weakly dominated by $(m_1, \ldots, m_{k-1}, m_k^*, m_{k+1}^*, \ldots, m_p^*)$. Against any of Player 2's *remaining strategies*, the latter gives a better payoff if play reaches node c_k, and the same payoff if it does not.

3. Backward induction or iterated elimination of weakly dominated strategies eventually produces the unique strategies $(m_1^*, \ldots, m_p^*)$ for Player 1 and $(n_1^*, \ldots, n_q^*)$ for Player 2.

2.13   Critique of elimination of dominated strategies

We saw with the Ultimatum Game in Section 1.7 that long backward inductions can lead to strange conclusions, and we explained in Section 1.13 that rational players need not follow the recommendation of a long backward induction. Iterating elimination of dominated strategies many times is subject to the same remarks: it can lead to strange conclusions, and rational players need not follow where it leads.

In addition, we have seen that iterated elimination of weakly dominated strategies can lead to different reduced games when done in different orders. We will encounter another issue with iterated elimination of weakly dominated strategies in the following chapter (see Section 3.6).

2.14   Problems

2.14.1   The Tragedy of the Commons.   There are n herders who share a common pasture. The pasture can support mn cattle without degrading. The ith herder has a choice of two strategies:

- The responsible strategy: graze m cattle.
- The irresponsible strategy: graze $m + 1$ cattle.

Each cow that is grazed brings a profit $p > 0$ to its herder. However, each herder who grazes $m + 1$ cattle imposes a cost $c > 0$ on the community of herders because of the degradation of the pasture. The cost is shared equally by the n herders.

Assume $c/n < p < c$. Thus the cost to the community of grazing an extra cow is greater than the profit from the cow, but each herder's share of the cost is less than the profit.

(1) Show that for each herder, grazing $m+1$ cattle strictly dominates grazing m cattle.
(2) Which of the following gives a higher payoff to each herder? (i) Every herder grazes $m + 1$ cattle. (ii) Every herder grazes m cattle. Give the payoffs in each case.

2.14.2 Another Auction.   A Ming vase is sold at auction. The auction works like this. Every bidder raises her hand. The auctioneer than calls out 1 dollar. Every bidder who is not willing to pay this price lowers her hand. If no hands remain up, the auction is over, and the vase is not sold to anyone. If exactly one buyer still has her hand up, the vase is sold to her for 1 dollar. If more than one buyer still has her hand up, the auctioneer calls out 2 dollars.

The auction continues in this manner. Once a bidder lowers her hand, she cannot raise it again. All the bidders decide simultaneously whether to lower their hands.

Notice that the auction ends when either (i) the auctioneer calls out k dollars and all hands are lowered but one, or (ii) the auctioneer calls out k dollars and all hands are lowered. In the first case, the vase is sold to the remaining bidder for k dollars. In the second case, the vase is not sold to anyone.

There are n bidders, $n \geq 2$. The value of the vase to Bidder i is v_i dollars; v_i is a nonnegative integer (0, 1, 2, ...). The payoff to Bidder i is 0 if Bidder i does not win the vase, and is v_i minus the price if Bidder i does win the vase.

A strategy for Bidder i is simply the highest bid she is willing to make, which we assume is a nonnegative integer.

(1) Suppose Bidder i's strategy is b_i, and the highest bid any other bidder is willing to make is h_i. Explain the following sentence: Bidder i's payoff π_i is determined by the two numbers b_i and h_i, and

$$\pi_i(b_i, h_i) = \begin{cases} 0 & \text{if } b_i \leq h_i, \\ v_i - (h_i + 1) & \text{if } h_i < b_i. \end{cases}$$

(2) Show that Bidder i's strategies $v_i - 1$ and v_i weakly dominate all her other strategies b_i with $b_i < v_i - 1$. To show this, pick a strategy $b_i < v_i - 1$ for Bidder i. Let h_i be the highest bid any other bidder is willing to make. Compare $\pi_i(v_i - 1, h_i)$ and $\pi_i(v_i, h_i)$ to $\pi_i(b_i, h_i)$. You will have to consider separately the three cases $h_i < b_i$, $b_i \leq h_i < v_i - 1$, and $v_i - 1 \leq h_i$.

(3) Show that Bidder i's strategies $v_i - 1$ and v_i weakly dominate all her other strategies b_i with $b_i > v_i$. To show this, pick a strategy $b_i > v_i$ for Bidder i. Let h_i be the highest bid any other bidder is willing to make. Compare $\pi_i(v_i - 1, h_i)$ and $\pi_i(v_i, h_i)$ to $\pi_i(b_i, h_i)$. You will have to consider separately the four cases $h_i < v_i - 1$, $h_i = v_i - 1$, $v_i \leqslant h_i < b_i$, and $b_i \leqslant h_i$.

(4) We conclude that Bidder i's strategies $v_i - 1$ and v_i weakly dominate all other strategies of Bidder i. Does either of Bidder i's strategies $v_i - 1$ and v_i weakly dominate the other? Explain.

2.14.3 Practice on iterated elimination of dominated strategies.

Use iterated elimination of dominated strategies to reduce the following games to smaller games in which iterated elimination of dominated strategies can no longer be used. State the order in which you eliminate strategies, and, for each strategy that you eliminate, state which remaining strategy dominates it. If you find a dominant strategy equilibrium, state what it is. (You eliminate rows by comparing first entries in the two rows, and you eliminate columns by comparing second entries in the two columns.)

(1) In this problem, use iterated elimination of strictly dominated strategies:

	t_1	t_2	t_3
s_1	$(73, 25)$	$(57, 42)$	$(66, 32)$
s_2	$(80, 26)$	$(35, 12)$	$(32, 54)$
s_3	$(28, 27)$	$(63, 31)$	$(54, 29)$

(2) In this problem, use iterated elimination of weakly dominated strategies:

	t_1	t_2	t_3	t_4	t_5
s_1	$(63, 1)$	$(28, 20)$	$(-2, 0)$	$(-2, 45)$	$(-3, 19)$
s_2	$(32, 3)$	$(2, 0)$	$(2, 5)$	$(33, 0)$	$(2, 3)$
s_3	$(54, -2)$	$(95, 4)$	$(0, 2)$	$(4, -1)$	$(0, 4)$
s_4	$(1, 33)$	$(-3, 43)$	$(-1, 39)$	$(1, -12)$	$(1, 17)$
s_5	$(54, 0)$	$(1, -13)$	$(-1, 88)$	$(-2, -57)$	$(-3, -1)$

2.14.4 War between Two Cities.

Cities A and I are at war. They are connected by a network of roads, each of which takes one day to travel; see Figure 2.1. City I sends out an army with many siege guns and four days' worth of supplies, and at the same time city A sends out a lightly armed force to stop city I's army. If city I's army arrives at city A after four days, it will use its siege guns to destroy city A's walls and win the war.

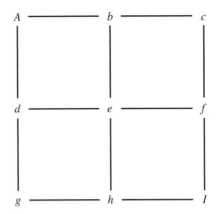

Figure 2.1. Cities and roads.

Each night for three nights, city I's army arrives at a road junction where it spends the night. If city A's force manages to arrive at the same junction on the same night, it will be able to attack city I's sleeping army and win the war.

Each country's strategy is a sequence of three road junctions where its army arrives on three successive nights. City I's army must stay at b or d on the third night. We will assume that city A's army must move every day.

(1) List city I's strategies. (There are six.)
(2) Explain why each strategy of city A that includes a visit to A is weakly dominated by a strategy that never visits A. (Hint: the strategy bAd is weakly dominated by the strategy bed: they do equally well on the first and third nights, but the second strategy will sometimes win on the second night, while the first strategy never will.)
(3) Explain why each strategy of city A that includes a visit to f or h is weakly dominated by a strategy that never visits f or h.
(4) List the remaining strategies for city A after these weakly dominated strategies have been eliminated. (There are six.)
(5) Let city I be Player 1 and let city A be Player 2. Construct a 6×6 payoff matrix that represents this game when each player is limited to its remaining six strategies. Use payoffs 1 to the winner and -1 to the loser.
(6) By eliminating weakly dominated strategies, reduce your game to a 2×4 one.

2.14.5 Football. On a play in a football game, the offense has two strategies, run and pass. The defense has three strategies, counter run, counter pass, and blitz. Payoff to the offense is yards gained; payoff to the defense is minus this number. Statistics indicate that the payoffs are given by the following matrix:

		Defense		
		counter run	counter pass	blitz
Offense	**run**	$(3, -3)$	$(6, -6)$	$(15, -15)$
	pass	$(10, -10)$	$(7, -7)$	$(9, -9)$

Use iterated elimination of strictly dominated strategies to find a dominant strategy equilibrium.

2.14.6 The Traveler's Dilemma. Two salesmen make identical trips in their cars. They both cross five bridges. Some of the bridges have one dollar tolls. Their boss knows that two of the bridges have one dollar tolls, but he doesn't know whether the others have one dollar tolls or are free. He asks each salesman to separately report his bridge expenses, which will be an integer between 2 and 5. As an incentive to report accurately, the boss makes the following rule. If both salesmen report the same amount, both will be reimbursed that amount. However, if they report different amounts, the boss will assume the lower amount is correct. He will reimburse the salesman who reported the lower amount that amount plus a $2 reward for honesty, and he will reimburse the other salesman that amount minus a $2 penalty for trying to cheat. The payoffs are therefore given by the following table:

		Salesman 2			
		2	3	4	5
	2	$(2, 2)$	$(4, 0)$	$(4, 0)$	$(4, 0)$
Salesman 1	3	$(0, 4)$	$(3, 3)$	$(5, 1)$	$(5, 1)$
	4	$(0, 4)$	$(1, 5)$	$(4, 4)$	$(6, 2)$
	5	$(0, 4)$	$(1, 5)$	$(2, 6)$	$(5, 5)$

Use iterated elimination of weakly dominated strategies to find a dominant strategy equilibrium.

2.14.7 Trying to Be Below Average. n players each independently choose an integer between 0 and 100. Let k denote the average of the chosen numbers. The player whose chosen number is closest to $\frac{2}{3}k$ wins a prize. If several players are closest to the chosen number, they share the prize equally.

(1) Explain the following statement: for each player, the choice 67 weakly dominates all choices greater than 67. Hint: what is the greatest $\frac{2}{3}k$ can possibly be?

(2) Eliminate all choices greater than 67 for all players. Justify the following statement: in the reduced game, for each player the choice 45 weakly dominates all choices greater than 45.

(3) Proceeding in this manner, reduce the game to one in which each player has just two strategies, 0 and 1. Can the game be reduced further?

Chapter 3

Nash equilibria

In this chapter we continue to look at games in normal form. We define the Nash equilibrium, which is a strategy profile at which no player, acting alone, can make a change that improves his outcome. Remarkably, this concept in its modern generality only dates back to around 1950, when John Nash developed it during his graduate work in mathematics at Princeton University. The earliest version of the Nash equilibrium occurs in Cournot's 1838 work on duopoly, which we discuss in Section 3.11. Nash equilbria are strategy profiles at which a game can get stuck. They may be good outcomes for all, some, or none of the players.

3.1 Big Monkey and Little Monkey 3 and the definition of Nash equilibria

Many games in normal form cannot be analyzed by elimination of dominated strategies. For example, in the encounter between Big Monkey and Little Monkey (see Section 1.5), suppose Big Monkey and Little Monkey decide simultaneously whether to wait or climb. Then we get the following payoff matrix:

		Little Monkey	
		wait	climb
Big Monkey	wait	$(0,0)$	$(9,1)$
	climb	$(4,4)$	$(5,3)$

There are no dominated strategies.

Consider a game in normal form with n players, strategy sets $S_1, \ldots, S_n$, and payoff functions $\pi_1, \ldots, \pi_n$. A *Nash equilibrium* is a strategy profile $(s_1^*, \ldots, s_n^*)$ with the following property: if any single player changes his strategy, his own payoff will not increase.

In other words, $(s_1^*, \ldots, s_n^*)$ is a Nash equilibrium provided

- For every $s_1 \in S_1$, $\pi_1(s_1^*, s_2^*, \ldots, s_n^*) \geqslant \pi_1(s_1, s_2^*, \ldots, s_n^*)$.
- For every $s_2 \in S_2$, $\pi_2(s_1^*, s_2^*, s_3^*, \ldots, s_n^*) \geqslant \pi_2(s_1^*, s_2, s_3^*, \ldots, s_n^*)$.
$\vdots$
- For every $s_n \in S_n$, $\pi_n(s_1^*, \ldots, s_{n-1}^*, s_n^*) \geqslant \pi_n(s_1^*, \ldots, s_{n-1}^*, s_n)$.

A *strict* Nash equilibrium is a strategy profile $(s_1^*, \ldots, s_n^*)$ with the property: if any single player changes his strategy, his own payoff will decrease.

In other words, $(s_1^*, \ldots, s_n^*)$ is a strict Nash equilibrium provided

- For every $s_1 \neq s_1^*$ in S_1, $\pi_1(s_1^*, s_2^*, \ldots, s_n^*) > \pi_1(s_1, s_2^*, \ldots, s_n^*)$.
- For every $s_2 \neq s_2^*$ in S_2, $\pi_2(s_1^*, s_2^*, s_3^*, \ldots, s_n^*) > \pi_2(s_1^*, s_2, s_3^*, \ldots, s_n^*)$.
$\vdots$
- For every $s_n \neq s_n^*$ in S_n, $\pi_n(s_1^*, \ldots, s_{n-1}^*, s_n^*) > \pi_n(s_1^*, \ldots, s_{n-1}^*, s_n)$.

In the game of Big Monkey and Little Monkey described above, there are two strict Nash equilibria:

- The strategy profile (wait, climb) is a strict Nash equilibrium. It produces the payoffs $(9, 1)$. If Big Monkey changes to climb, his payoff decreases from 9 to 5. If Little Monkey changes to wait, his payoff decreases from 1 to 0.
- The strategy profile (climb, wait) is also a strict Nash equilibrium. It produces the payoffs $(4, 4)$. If Big Monkey changes to wait, his payoff decreases from 4 to 0. If Little Monkey changes to climb, his payoff decreases from 4 to 3.

Big Monkey prefers the first of these Nash equilibria, Little Monkey the second.

Note that in this game, the strategy profiles (wait, wait) and (climb, climb) are not Nash equilibria. In fact, for these strategy profiles, either monkey could improve his payoff by changing his strategy.

Game theorists often use the following notation when discussing Nash equilibria. Let $s = (s_1, \ldots, s_n)$ denote a strategy profile. Suppose in s we replace the ith player's strategy s_i by another of his strategies, say, s_i'. The resulting strategy profile is then denoted (s_i', s_{-i}).

In this notation, a strategy profile $s^* = (s_1^*, \ldots, s_n^*)$ is a Nash equilibrium if, for each $i = 1, \ldots, n$, $\pi_i(s^*) \geqslant \pi_i(s_i, s_{-i}^*)$ for every $s_i \in S_i$. The strategy profile s^* is a strict Nash equilibrium if, for each $i = 1, \ldots, n$, $\pi_i(s^*) > \pi_i(s_i, s_{-i}^*)$ for every $s_i \neq s_i^*$ in S_i.

The notion of Nash equilibrium is the most important idea in game theory. We consider three ways of finding Nash equilibria:

- Inspection (Sections 3.2–3.5).
- Iterated elimination of dominated strategies (Sections 3.6–3.7).
- Using best response (Sections 3.8–3.11).

3.2 Finding Nash equilibria by inspection: Important examples

One way to find Nash equilibria is by inspection of all strategy profiles to see which, if any, meet the definition. This is how we found the Nash equilibria in the game of Big Monkey and Little Monkey in the previous section. Here are four important examples.

3.2.1 Prisoner's Dilemma. Recall the Prisoner's Dilemma from Section 2.1:

		Executive 2	
		refuse	talk
Executive 1	refuse	$(-1, -1)$	$(-10, 0)$
	talk	$(0, -10)$	$(-6, -6)$

Let's inspect all four strategy profiles:

- The strategy profile (refuse, refuse) is not a Nash equilibrium. If either executive alone changes his strategy to talk, his payoff increases from -1 to 0.
- The strategy profile (refuse, talk) is not a Nash equilibrium. If Executive 1 changes his strategy to talk, his payoff increases from -10 to -6.
- The strategy profile (talk, refuse) is not a Nash equilibrium. If Executive 2 changes his strategy to talk, his payoff increases from -10 to -6.
- The strategy profile (talk, talk) is a strict Nash equilibrium. If either executive alone changes his strategy to refuse, his payoff falls from -6 to -10.

The Prisoner's Dilemma illustrates an important fact about Nash equilibria: they are not necessarily good for the players! Instead they are strategy profiles where a game can get stuck, for better or worse.

3.2.2 Stag Hunt. Two hunters on horseback are pursuing a stag. If both work together, they will succeed. However, the hunters notice that they are passing some hares. If either hunter leaves the pursuit of the stag to pursue a hare, he will succeed, but the stag will escape.

Let's model this situation as a two-player game in which the players decide their moves simultaneously. The players are the hunters. The possible strategies for each are pursue the stag and pursue a hare. Let's suppose that the payoff for catching a hare is 1 to the hunter who caught it, and the payoff for catching the stag is 2 to each hunter. The payoff matrix is

		Hunter 2	
		stag	hare
Hunter 1	stag	$(2,2)$	$(0,1)$
	hare	$(1,0)$	$(1,1)$

There are no dominated strategies. If we inspect all four strategy profiles, we find that there are two strict Nash equilibria, (stag, stag) and (hare, hare). Both hunters prefer (stag, stag) to (hare, hare).

Like the Prisoner's Dilemma, Stag Hunt represents a type of cooperation dilemma that is common in human affairs. Without cooperating, each hunter can decide individually to pursue a hare, thus guaranteeing himself a payoff of 1. If both players do this, we have a noncooperative Nash equilibrium. A better Nash equilibrium exists in which both players cooperate to pursue the stag. However, if the players are in the noncooperative equilibrium, it may be difficult to get to the cooperative equilibrium. The reason is that if one player on his own switches to the strategy of pursuing the stag, his payoff becomes worse.

Stag Hunt differs from the Prisoner's Dilemma in that, if both players manage to cooperate by pursuing the stag together, they have arrived at a Nash equilibrium, which means that neither will be tempted to cheat.

The problem of how to deal with a depressed economy resembles a Stag Hunt game. In a depressed economy, companies don't hire workers because they lack customers, and they lack customers because other companies don't hire workers. This is a low-payoff equilibrium. A better equilibrium is one in which lots of companies hire lots of workers, who are then needed because the other companies' workers become customers. However, it is not easy to get from the first equilibrium to the second. The first companies to hire more workers will find that the new workers are not needed and depress profits.

The solution to this problem proposed by John Maynard Keynes in the 1930s is that in a depressed economy, the government should spend more, for example, on infrastructure projects. The additional spending will cause some companies to hire more workers, who will become customers for other companies, which will in turn hire more workers. Once the economy has been jolted into the high-payoff equilibrium, it can maintain itself there.

Stag Hunt problems, while difficult to solve, are easier to solve than Prisoner's Dilemmas. Thus the problem of a depressed economy should be easier to solve than the problem of global warming; compare Section 2.5.

The Wikipedia page for Stag Hunt is http://en.wikipedia.org/wiki/Stag_hunt. Games like Stag Hunt, in which there are several Nash equilibria, one of which is best for all players, are sometimes called *pure coordination games*. The players should somehow coordinate their actions so that they are in the best of the Nash equilibria. For more information, see the Wikipedia page http://en.wikipedia.org/wiki/Coordination_game.

3.2.3 Chicken. The game of Chicken was (supposedly) played by American teenagers in the 1950s. A variant of the game (not the version we describe) is shown in the James Dean movie "Rebel without a Cause." In Chicken, two teenagers drive their cars toward each other at high speed. Each has two strategies: drive straight or swerve. The payoffs are as follows:

		Teenager 2	
		straight	swerve
Teenager 1	straight	$(-2, -2)$	$(1, -1)$
	swerve	$(-1, 1)$	$(0, 0)$

If one teen drives straight and one swerves, the one who drives straight gains in reputation, and the other loses face. However, if both drive straight, there is a crash, and both are injured. There are two Nash equilibria: (straight, swerve) and (swerve, straight). Each gives a payoff of 1 to one player and -1 to the other. Thus each player prefers the equilibrium that gives the payoff 1 to himself.

The version of Big Monkey and Little Monkey discussed in Section 3.1 is a Chicken-type game.

One way to win Chicken-type games is to cultivate a reputation for being crazy, and hence willing to pursue the drive straight strategy even though it may lead to disaster. For a scary example of how President Richard Nixon and his National Security Advisor Henry Kissinger used this idea during negotiations to end the Vietnam War, see http://www.wired.com/politics/security/magazine/16-03/ff_nuclearwar.

3.2.4 Battle of the Sexes. Alice and Bob want to meet this evening. There are two events they could meet at: a Justin Bieber concert and a pro wrestling match. Unfortunately, their cell phones are dead. Alice prefers the concert, and Bob prefers the wrestling match. However, they both prefer meeting to missing each other. The payoffs are given in the following table.

		Bob	
		concert	wrestling
Alice	concert	$(2,1)$	$(0,0)$
	wrestling	$(0,0)$	$(1,2)$

As in Chicken, there are two Nash equilibria, one preferred by one player, one by the other. Here, however, the players are trying to cooperate rather than compete. If you were Alice, which event would you go to?

3.3   Water Pollution 1

Three firms use water from a lake. When a firm returns the water to the lake, it can purify it or fail to purify it (and thereby pollute the lake). The cost of purifying the used water before returning it to the lake is 1. If two or more firms fail to purify the water before returning it to the lake, all three firms incur a cost of 3 to treat the water before they can use it.

The payoffs are therefore as follows:

- If all three firms purify: -1 to each firm.
- If two firms purify and one pollutes: -1 to each firm that purifies, 0 to the polluter.
- If one firm purifies and two pollute: -4 to the firm that purifies, -3 to each polluter.
- If all three firms pollute: -3 to each firm.

We inspect these possibilities to see whether any are Nash equilibria:

- Suppose all three firms purify. If one switches to polluting, its payoff increases from -1 to 0. This is not a Nash equilibrium.
- Suppose two firms purify and one pollutes. If a purifier switches to polluting, its payoff decreases from -1 to -3. If the polluter switches to purifying, its payoff decreases from 0 to -1. Thus there are three Nash equilibria in which two firms purify and one pollutes.
- Suppose one firm purifies and two pollute. If the purifier switches to polluting, its payoff increases from -4 to -3. This is not a Nash equilibrium.
- Suppose all three firms pollute. If one switches to purifying, its payoff decreases from -3 to -4. This is a Nash equilibrium.

We see here an example of the *free rider problem* (Wikipedia page: http://en.wikipedia.org/wiki/Free_rider_problem). Each firm wants to be the one that gets the advantage of the other firms' efforts without making any effort itself.

The free rider problem arises in negotiating treaties to deal with climate change. For example, the United States objected to the 1997 Kyoto protocol, because it did not require action by developing countries, such as China and India (Wikipedia page: http://en.wikipedia.org/wiki/Kyoto_treaty).

The free rider problem also arose in connection with the Troubled Asset Relief Program (TARP), under which the U.S. Treasury invested several hundred billion dollars in U.S. banks during the financial crisis of late 2008 and early 2009. (Wikipedia page: http://en.wikipedia.org/wiki/Troubled_Asset_Relief_Program.) Some banks wanted to pay back this investment very quickly. What could be wrong with this?

- The banks had made many loans to people and corporations who appeared unable to repay them. If too many of these loans were not repaid, the banks would be rendered bankrupt.
- The banks therefore needed to conserve what cash they had, so they were unwilling to make new loans.
- If the banks were unwilling to lend, the economy would slow, making it even less likely that that the problem loans would be repaid.
- The government therefore invested in (injected capital into) the banks. The banks now had more money, so they would, the government hoped, lend more. Then the economy would pick up, some of the problem loans would be repaid, and the banks would be okay.
- Unfortunately for the banks, the government's investment was accompanied by annoying requirements, such as limitations on executive pay.
- If a few banks were allowed to repay the government's investment, they could avoid the annoying requirements, but still benefit from the economic boost and loan repayments due to other banks' increased lending. The banks that repaid the government would become free riders.

3.4 Arguing over Marbles

Two children begin to argue about some marbles with a value of 1. If one child gives up arguing first, the other child gets the marbles. If both children give up arguing at the same time, they split the marbles.

The payoff to each child is the value of the marbles she gets, minus the length of time in hours that the argument lasts.

After one hour, it will be time for dinner. If the argument has not ended before then, it ends then, and the children split the marbles.

We will treat this situation as a two-player game. Before the game begins, each child decides independently how long she is willing to argue, in hours. We will allow this choice to be any real number between 0 and 1. Thus the first child's strategy is a real number s, $0 \leqslant s \leqslant 1$, and the second child's strategy is a real number t, $0 \leqslant t \leqslant 1$.

The payoffs are:

- If $s < t$, the argument ends after s hours, and the second child gets the marbles, so $\pi_1(s,t) = -s$ and $\pi_2(s,t) = 1 - s$.
- If $s > t$, the argument ends after t hours and the first child gets the marbles, so $\pi_1(s,t) = 1 - t$ and $\pi_2(s,t) = -t$.
- If $s = t$, the argument ends after s hours and the children split the marbles, so $\pi_1(s,t) = \frac{1}{2} - s$ and $\pi_2(s,t) = \frac{1}{2} - s$.

We will find the Nash equilibria by inspecting all strategy profiles (s,t), $0 \leqslant s \leqslant 1, 0 \leqslant t \leqslant 1$.

We first inspect strategy profiles (s,t) with $s < t$.

1. Suppose $0 < s < t \leqslant 1$. Player 2's payoff is $1 - s$, and Player 1's is $-s$. Player 1 could improve her payoff by reducing s. None of these profiles (s,t) is a Nash equilibrium.

2. Suppose $s = 0$ and $0 < t < 1$. Player 2's payoff is 1, and Player 1's payoff is 0. Player 1 could improve her payoff by changing to a number s' with $t < s' \leqslant 1$. Player 1 would then win the argument, so her payoff would increase to $1 - t$, which is greater than 0. (This also works in the previous case, provided $t < 1$.) None of these profiles $(0,t)$ is a Nash equilibrium.

3. The only remaining possibility with $s < t$ is $s = 0$ and $t = 1$. Player 2's payoff is 1, and Player 1's payoff is 0. Neither player can improve her payoff by changing her strategy. This is a Nash equilibrium.

Thus the only Nash equilibrium with $s < t$ is $(s,t) = (0,1)$. Similarly, the only Nash equilibrium with $s > t$ is $(s,t) = (1,0)$.

Next we inspect strategy profiles (s,t) with $s = t$.

1. Suppose $s = t < 1$. Both players get the payoff $\frac{1}{2} - s$. Player 1, for example, could improve her payoff by changing to a number s' with $s < s' \leqslant 1$. Player 1 would then win the argument, so her payoff would increase from $\frac{1}{2} - s$ to $1 - s$. None of these profiles is a Nash equilibrium.

2. Suppose $s = t = 1$. Both players get the payoff $-\frac{1}{2}$. Either player could improve her payoff by changing to 0, thus losing the argument and getting a payoff of 0. Therefore $(s,t) = (1,1)$ is not a Nash equilibrium.

We conclude that the only Nash equilibria are $(0,1)$ and $(1,0)$. The first is better for Player 2, the second for Player 1. This game is reminiscent of Chicken. If one player is willing to argue for the entire hour, the other player can do no better than to give up and let her have all the marbles.

Arguing over Marbles is a variant of the well-known game War of Attrition. In War of Attrition, the two players are interpreted as contestants competing for a resource of value 1; the numbers s and t represent the resources that the contestants are willing to commit to the competition. See the Wikipedia page http://en.wikipedia.org/wiki/War_of_attrition_(game) for more information.

3.5 Tobacco Market

At a certain warehouse, the price of tobacco per pound in dollars, p, is related to the supply of tobacco in pounds, q, by the formula

$$p = 10 - \frac{q}{100{,}000}. \tag{3.1}$$

Thus the more tobacco that farmers bring to the warehouse, the lower the price becomes. However, a price support program ensures that the price never falls below \$.25 per pound. In other words, if the supply is so high that the price would be below \$.25 per pound, the price is set at $p = .25$, and a government agency purchases whatever cannot be sold at that price.

One day three farmers are the only ones bringing their tobacco to this warehouse. Each has harvested 600,000 pounds and can bring as much of his harvest as he wants. Whatever is not brought must be discarded.

There are three players, the farmers. Farmer i's strategy is simply the amount of tobacco he brings to the warehouse, and hence is a number q_i, $0 \leqslant q_i \leqslant 600{,}000$. The payoff to Farmer i is $\pi_i(q_1, q_2, q_3) = pq_i$, where

$$p = \begin{cases} 10 - \dfrac{q_1 + q_2 + q_3}{100{,}000} & \text{if } q_1 + q_2 + q_3 \leqslant 975{,}000, \\ .25 & \text{if } q_1 + q_2 + q_3 > 975{,}000. \end{cases}$$

(The importance of 975,000 is that if $q = 975{,}000$, then (3.1) gives $p = .25$.) We find the Nash equilibria by inspecting all strategy profiles (q_1, q_2, q_3), $0 \leqslant q_i \leqslant 600{,}000$.

1. Suppose some $q_i = 0$. Then Farmer i's payoff is 0, which he could increase by bringing some of his tobacco to market. This is not a Nash equilibrium.

2. Suppose $q_1 + q_2 + q_3 \geqslant 975{,}000$. The price is then \$.25, and will stay the same if any farmer brings more of his tobacco to market. Thus if any q_i is less than 600,000, that farmer could increase his own payoff by bringing more of his tobacco to market. Hence the only possible Nash equilibrium with $q_1 + q_2 + q_3 \geqslant 975{,}000$ is (600,000, 600,000, 600,000). It really is one: if any farmer alone brings less to market, the price will not rise, so his payoff will certainly decrease. The payoff to each farmer at this Nash equilibrium is $\pi_i = .25 \times 600{,}000 = 150{,}000$.

3. Suppose $q_1 + q_2 + q_3 < 975{,}000$ and $0 < q_i < 600{,}000$ for all i. In this region the payoff functions π_i are given by

$$\pi_i(q_1, q_2, q_3) = \left(10 - \frac{q_1 + q_2 + q_3}{100{,}000}\right) q_i.$$

Suppose (q_1, q_2, q_3) is a Nash equilibrium in this region. Let us consider first Farmer 1. The maximum value of $\pi_1(q_1, q_2, q_3)$, with q_2 and q_3 fixed at their Nash equilibrium values, must occur at the Nash equilibrium. Since q_1 is not an endpoint of the interval $0 \leqslant q_1 \leqslant 600,000$, we must have $\partial \pi_1 / \partial q_1 = 0$ at the Nash equilibrium. By considering Farmers 2 and 3, we get the additional equations $\partial \pi_2 / \partial q_2 = 0$ and $\partial \pi_3 / \partial q_3 = 0$. This is a system of three equations in the three unknowns (q_1, q_2, q_3). If you solve it, you will find the only possible Nash equilibrium in the region under consideration.

Our system of equations is

$$10 - \frac{q_1 + q_2 + q_3}{100,000} - \frac{q_1}{100,000} = 0,$$

$$10 - \frac{q_1 + q_2 + q_3}{100,000} - \frac{q_2}{100,000} = 0,$$

$$10 - \frac{q_1 + q_2 + q_3}{100,000} - \frac{q_3}{100,000} = 0.$$

It is a linear system. You can solve it with a calculator, or you can write it in a standard form and use row operations to reduce it to row-echelon form. Here is another way. The three equations imply that $q_1 = q_2 = q_3$. (You can see this by solving the ith equation for $q_i/100,000$. You get the same answer for each i.) Then the first equation implies that

$$10 - \frac{3q_1}{100,000} - \frac{q_1}{100,000} = 0,$$

so $q_1 = 250,000$. Hence the only possible Nash equilibrium in this region is $(250,000, 250,000, 250,000)$.

To check whether this really is a Nash equilibrium, let us consider Farmer 1. (The others are similar). For $(q_2, q_3) = (250,000, 250,000)$, Farmer 1's payoff function is

$$\pi_1(q_1, 250,000, 250,000)$$
$$= \begin{cases} \left(10 - \dfrac{q_1 + 500,000}{100,000}\right) q_1 & \text{if } 0 \leqslant q_1 \leqslant 475,000, \\ .25q_1 & \text{if } 475,000 < q_1 \leqslant 600,000. \end{cases}$$

The quadratic function $(10 - (q_1 + 500,000)/100,000) q_1$, $0 \leqslant q_1 \leqslant 475,000$, is maximum at the point we have found, $q_1 = 250,000$, where $\pi_1 = 2.50 \times 250,000 = 625,000$. Moreover, for $475,000 < q_1 \leqslant 600,000$, $\pi_1 = .25q_1$ is at most 150,000. Therefore Farmer 1 cannot improve his payoff by changing q_1. The same is true for Farmers 2 and 3, so $(250,000, 250,000, 250,000)$ is indeed a Nash equilibrium.

4. There is one case we have not yet considered: $q_1 + q_2 + q_3 < 975{,}000$, $0 < q_i < 600{,}000$ for two i, and $q_i = 600{,}000$ for one i. It turns out that there are no Nash equilibria in this case. The analysis is left as homework.

In conclusion, there are two Nash equilibria, (600,000, 600,000, 600,000) and (250,000, 250,000, 250,000). The second is preferred to the first by all three farmers. Therefore Tobacco Market is a pure coordination game, like Stag Hunt (Subsection 3.2.2). If the farmers can agree among themselves to each bring 250,000 pounds of tobacco to market and discard 350,000 pounds, none will have an incentive to cheat. However, the tobacco buyers would prefer that they each bring 600,000 pounds to market. If all farmers do that, none can improve his own payoff be bringing less. Thus, as with Stag Hunt, if the farmers are in the equilibrium in which they each bring all their tobacco to market, it may be difficult for them to get to the other equilibrium.

3.6 Finding Nash equilibria by iterated elimination of dominated strategies

The relation between iterated elimination of dominated strategies, which we discussed in Chapter 2, and Nash equilibria is summarized in the following theorems.

Theorem 3.1. *Suppose we do iterated elimination of* weakly *dominated strategies on a game G in normal form. Let H be the reduced game that results. Then:*

(1) Each Nash equilibrium of H is also a Nash equilibrium of G.
(2) In particular, if H has only one strategy s_i^* for each player, then the strategy profile $(s_1^*, \ldots, s_n^*)$ is a Nash equilibrium of G.

The last conclusion of Theorem 3.1 just says that every dominant strategy equilibrium is a Nash equilibrium.

For iterated elimination of *strictly* dominated strategies, one can say more.

Theorem 3.2. *Suppose we do iterated elimination of* strictly *dominated strategies on a game G in normal form. Then:*

(1) Any order yields the same reduced game H.
(2) Each strategy that is eliminated is not part of any Nash equilibrium of G.
(3) Each Nash equilibrium of H is also a Nash equilibrium of G.
(4) If H has only one strategy s_i^* for each player, then $(s_1^*, \ldots, s_n^*)$ is a strict Nash equilibrium. Moreover, there are no other Nash equilibria.

Theorem 3.2 justifies using iterated elimination of strictly dominated strategies to reduce the size of the game to be analyzed. It says in part that we do not miss any Nash equilibria by doing the reduction.

In contrast, we certainly can miss Nash equilibria by using iterated elimination of weakly dominated strategies to reduce the size of the game. This is the second problem with iterated elimination of weakly dominated strategies that we have identified; the first was that the resulting smaller game can depend on the order in which the elimination is done.

We shall prove statements (2), (3), and (4) of Theorem 3.2, and we indicate how our proof of statement (3) of Theorem 3.2 can be modified to give a proof of statement (1) of Theorem 3.1. Of course, statement (2) of Theorem 3.1 follows from statement (1) of Theorem 3.1.

Proof. We consider iterated elimination of strictly dominated strategies on a game G in normal form.

To prove statement (2) of Theorem 3.2, we prove by induction the statement: the kth strategy that is eliminated is not part of any Nash equilibrium.

$k = 1$: Let s_i, a strategy of Player i, be the first strategy that is eliminated. It was eliminated because the ith player has a strategy t_i that strictly dominates it. Suppose s_i is part of a strategy profile. Replacing s_i by t_i, and leaving all other players' strategies the same, increases the payoff to Player i. Therefore this strategy profile is not a Nash equilibrium.

Assume the statement is true for $k = 1, \ldots, l$. Let s_i, a strategy of Player i, be the $(l + 1)$st strategy that is eliminated. It was eliminated because the ith player has a strategy t_i that strictly dominates it, assuming no player uses any previously eliminated strategy. Suppose s_i is part of a strategy profile. If any of the previously eliminated strategies is used in this strategy profile, then by assumption it is not a Nash equilibrium. If none of the previously eliminated strategies is used, then replacing s_i by t_i, and leaving all other players' strategies the same increases the payoff to Player i. Therefore this strategy profile is not a Nash equilibrium.

This completes the proof of statement (2) of Theorem 3.2.

To prove statement (3) of Theorem 3.2, let $(s_1^*, s_2^*, \ldots, s_n^*)$ be a Nash equilibrium of H, and let S_1 be Player 1's strategy set for G. We must show that for every $s_1 \neq s_1^*$ in S_1,

$$\pi_1(s_1^*, s_2^*, \ldots, s_n^*) \geqslant \pi_1(s_1, s_2^*, \ldots, s_n^*). \tag{3.2}$$

(Of course, we must also prove an analogous statement for the other players, but the argument would be the same.)

If s_1 is a strategy of Player 1 that remains in the reduced game H, then of course (3.2) follows from the fact that $(s_1^*, s_2^*, \ldots, s_n^*)$ is a Nash equilibrium of H. To complete the proof, we show that if s_1 is any strategy of Player 1 that was eliminated in the course of iterated elimination of strictly dominated strategies, then

$$\pi_1(s_1^*, s_2^*, \ldots, s_n^*) > \pi_1(s_1, s_2^*, \ldots, s_n^*). \tag{3.3}$$

Note the strict inequality.

In fact, we prove by reverse induction the statement: if s_1 is the kth strategy of Player 1 to be eliminated, then (3.3) holds.

Let s_1 be the *last* strategy of Player 1 to be eliminated. It was eliminated because one of the remaining strategy of Player 1 (say, t_1), when used against any remaining strategies $s_2, \ldots, s_n$ of the other players, satisfied $\pi_1(t_1, s_2, \ldots, s_n) > \pi_1(s_1, s_2, \ldots, s_n)$. In particular, since $s_2^*, \ldots, s_n^*$ were among the remaining strategies, we get

$$\pi_1(t_1, s_2^*, \ldots, s_n^*) > \pi_1(s_1, s_2^*, \ldots, s_n^*). \tag{3.4}$$

Since $(s_1^*, s_2^*, \ldots, s_n^*)$ is a Nash equilibrium of H, and t_1 is a strategy available to Player 1 in H (it was never eliminated),

$$\pi_1(s_1^*, s_2^*, \ldots, s_n^*) \geqslant \pi_1(t_1, s_2^*, \ldots, s_n^*). \tag{3.5}$$

Combining (3.5) and (3.4), we get (3.3).

Assume the statement is true for all $k \geqslant \ell$. Let s_1 be the $(\ell-1)$st strategy of Player 1 to be eliminated. It was eliminated because one of the other remaining strategies of Player 1 (say, t_1), when used against any remaining strategies $s_2, \ldots, s_n$ of the other players, satisfied $\pi_1(t_1, s_2, \ldots, s_n) > \pi_1(s_1, s_2, \ldots, s_n)$. In particular, since $s_2^*, \ldots, s_n^*$ were among the remaining strategies, we get (3.4). If t_1, is never eliminated, then it is still available to Player 1 in H, so we have (3.5). Combining (3.5) and (3.4), we again get (3.3). If t_1 *is* one of the strategies of Player 1 that is eliminated after s_1, then, by the induction hypothesis,

$$\pi_1(s_1^*, s_2^*, \ldots, s_n^*) > \pi_1(t_1, s_2^*, \ldots, s_n^*). \tag{3.6}$$

Combining (3.6) and (3.4), we get (3.3).

This completes the proof of statement (3) of Theorem 3.2.

To prove statement (4) of Theorem 3.2, we must show that if H has only one strategy s_i^* for each player, then for every $s_1 \neq s_1^*$ in S_1,

$$\pi_1(s_1^*, s_2^*, \ldots, s_n^*) > \pi_1(s_1, s_2^*, \ldots, s_n^*). \tag{3.7}$$

In this case, every $s_1 \neq s_1^*$ in S_1 is eliminated in the course of iterated elimination of strictly dominated strategies, so the proof we gave of statement (3) actually yields the conclusion.

The proof of statement (1) of Theorem 3.1 is essentially the same as the proof of statement (3) of Theorem 3.2, except that no inequalities are strict.

□

3.7 Big Monkey and Little Monkey 4: Threats, promises, and commitments revisited

Let us consider again the game of Big Monkey and Little Monkey (Section 1.5) with Big Monkey going first. The normal form of this game, which was given in Section 2.10, is repeated here:

| | | \multicolumn{4}{c}{Little Monkey} |
		ww	wc	cw	cc
Big Monkey	w	$(0,0)$	$(0,0)$	$(9,1)$	$(9,1)$
	c	$(4,4)$	$(5,3)$	$(4,4)$	$(5,3)$

We have seen two ways of doing iterated elimination of weakly dominated strategies for this game.

One, which corresponds to a way of doing backward induction in the extensive form of the game, led to the 1×1 reduced game consisting of the strategy profile (w, cw). This strategy profile is therefore a dominant strategy equilibrium, and hence a Nash equilibrium. However, since it was found by iterated elimination of weakly dominated strategies, it is not guaranteed to be the only Nash equilibrium.

The second way we did iterated elimination of weakly dominated strategies led to a 2×1 reduced game. Both remaining strategy profiles (w, cw) (found before) and (w, cc) are Nash equilibria.

One can check that (c, ww) is also a Nash equilibrium. It cannot be found by doing iterated elimination of weakly dominated strategies.

Both Nash equilibria (w, cc) and (c, ww) use strategies that were eliminated in one way of doing iterated elimination of weakly dominated strategies, but they are strategies that we have seen before in Section 1.6. In the Nash equilibrium (c, ww), Little Monkey's strategy ww includes the threat that if Big Monkey waits, he will wait also. In the Nash equilibrium (w, cc), Little Monkey's strategy cc includes the promise to climb if Big Monkey climbs. You may recall that the threat changed the outcome of the game, but the promise did not.

Why do we find Nash equilibria when we look at the normal form of the game that we did not find using backward induction?

The reason is that when looking at the normal form of the game, *we assume that Big Monkey and Little Monkey choose their strategies once and for all at the start of the game.* The fact that Little Monkey's strategy might include a move that, should the time comes to make it, would not be profitable, is not relevant.

Consider again the strategy profile (c, ww), in which Little Monkey makes the threat to wait if Big Monkey waits. A little thought reveals that this is certainly a Nash equilibrium. If Little Monkey commits himself in advance to waiting no matter what, then Big Monkey can do no better than to climb. On the other hand, if Big Monkey climbs, Little Monkey can do no better than to wait, which he will indeed do if he adopts the strategy ww.

This analysis is relevant provided Little Monkey can really commit himself in advance to using the strategy ww, so that the normal form of the game becomes appropriate.

3.8 Finding Nash equilibria using best response

Consider a game in normal form with n players, strategy sets $S_1, \ldots, S_n$, and payoff functions $\pi_1, \ldots, \pi_n$. Let $s_2, \ldots, s_n$ be fixed strategies for players 2,...,n. Suppose s_1^* is a strategy for Player 1 with the property that

$$\pi_1(s_1^*, s_2, \ldots, s_n) \geqslant \pi_1(s_1, s_2, \ldots, s_n) \text{ for all } s_1 \in S_1. \tag{3.8}$$

Then s_1^* is a *best response* of Player 1 to the strategy choices $s_2, \ldots, s_n$ of the other players. Of course, Player 1 may have more than one such best response.

For each choice $s_2, \ldots, s_n$ of strategies by the other players, let $B_1(s_2, \ldots, s_n)$ denote the set of best responses by Player 1. In other words,

$$s_1^* \in B_1(s_2, \ldots, s_n) \text{ if and only if } \pi_1(s_1^*, s_2, \ldots, s_n) \geqslant \pi_1(s_1, s_2, \ldots, s_n)$$
$$\text{for all } s_1 \in S_1.$$

The mapping that associates to each $(s_2, \ldots, s_n) \in S_2 \times \cdots \times S_n$ the corresponding set $B_1(s_2, \ldots, s_n)$, a subset of S_1, is called Player 1's *best response correspondence*. If each set $B_1(s_2, \ldots, s_n)$ consists of a single strategy, we have Player 1's *best response function* $b_1(s_2, \ldots, s_n)$.

Best response correspondences for the other players are defined analogously.

At a Nash equilibrium, each player's strategy is a best response to the other players' strategies. In other words, the strategy profile $(s_1^*, \ldots, s_n^*)$ is a Nash equilibrium if and only if

- $s_1^* \in B_1(s_2^*, \ldots, s_n^*)$.
- $s_2^* \in B_2(s_1^*, s_3^* \ldots, s_n^*)$.
  $\vdots$
- $s_n^* \in B_n(s_1^*, \ldots, s_{n-1}^*)$.

This property of Nash equilibria can be used to find them. Just graph all players' best response correspondences in one copy of strategy profile space and find where they intersect! Alternatively, describe each best response correspondence by an equation, and solve the equations simultaneously.

3.9 Big Monkey and Little Monkey 5

Once again we consider the game of Big Monkey and Little Monkey (Sections 1.5 and 2.10) with Big Monkey going first. The normal form of this game with both players' best response correspondences graphed is shown here:

		Little Monkey			
		ww	*wc*	*cw*	*cc*
Big Monkey	*w*	$(0,0)$	$(0,0)$	$(\boxed{9},\boxed{1})$	$(\boxed{9},\boxed{1})$
	c	$(\boxed{4},\boxed{4})$	$(\boxed{5},3)$	$(4,\boxed{4})$	$(5,3)$

The explanation of the boxes is as follows:

- Big Monkey's best response correspondence is actually a function:
  - If Little Monkey does *ww*, do *c*.
  - If Little Monkey does *wc*, do *c*.
  - If Little Monkey does *cw*, do *w*.
  - If Little Monkey does *cc*, do *w*.

  This correspondence is indicated in the payoff matrix by drawing a box around the associated payoffs to Big Monkey. In other words, in each of the four columns of the matrix, we draw a box around the highest first entry, which is Big Monkey's highest payoff.
- Little Monkey's best response correspondence is not a function:
  - If Big Monkey does *w*, do *cw* or *cc*.
  - If Big Monkey does *c*, do *ww* or *cw*.

  This correspondence is indicated in the payoff matrix by drawing a box around the associated payoffs to Little Monkey. In other words, in each of the two rows of the matrix, we draw a box around the highest second entry, which is Little Monkey's highest payoff.

Notice that three ordered pairs have both payoffs boxed. These ordered pairs correspond to intersections of the graphs of the two players' best response correspondences, and hence to Nash equilibria.

For a two-player game in normal form where each player has only a finite number of strategies, graphing the best response correspondences as we did in this example is the best way to find Nash equilibria.

Table 3.1. Payoff matrices for the Water Pollution game

Firm 3 purifies

		Firm 2 purify	pollute
Firm 1	purify	$(-1,-1,-1)$	$(-1,0,-1)$
	pollute	$(0,-1,-1)$	$(-3,-3,-4)$

Firm 3 pollutes

		Firm 2 purify	pollute
Firm 1	purify	$(-1,-1,0)$	$(-4,-3,-3)$
	pollute	$(-3,-4,-3)$	$(-3,-3,-3)$

3.10 Water Pollution 2

The payoffs in the Water Pollution game (Section 3.3) can be represented by two 2×2 matrices of ordered triples as shown in Table 3.1. Each ordered triple represents payoffs to Firms 1, 2, and 3.

These two matrices should be thought of as stacked one above the other. We then indicate

- Player 1's best response to each choice of strategies by the other players by boxing the highest first entry in each column;
- Player 2's best response to each choice of strategies by the other players by boxing the highest second entry in each row; and
- Player 3's best response to each choice of strategies by the other players by boxing the highest third entry in each "stack."

The result is shown in Table 3.2. The Nash equilibria correspond to ordered triples with all three entries boxed. As before, we have found four Nash equilibria. In three of them, two firms purify and one pollutes. In the fourth, all firms pollute.

3.11 Cournot's model of duopoly

Cournot's model of duopoly (Wikipedia article: http://en.wikipedia.org/wiki/Cournot_duopoly) is the same as Stackelberg's (see Section 1.11), except that that the players choose their production levels simultaneously. This is a game in normal form with two players, strategy sets $0 \leqslant s < \infty$ and $0 \leqslant t < \infty$, and

Table 3.2. Payoff matrices in the Water Pollution game—best response

Firm 3 purifies

		Firm 2 purify	pollute
Firm 1	purify	$(-1, -1, -1)$	$(\boxed{-1}, \boxed{0}, \boxed{-1})$
	pollute	$(\boxed{0}, \boxed{-1}, \boxed{-1})$	$(-3, -3, -4)$

Firm 3 pollutes

		Firm 2 purify	pollute
Firm 1	purify	$(\boxed{-1}, \boxed{-1}, \boxed{0})$	$(-4, -3, -3)$
	pollute	$(-3, -4, -3)$	$(\boxed{-3}, \boxed{-3}, \boxed{-3})$

payoff functions

$$\pi_1(s, t) = ps - c_1(s) = \begin{cases} (\alpha - \beta(s+t))s - cs & \text{if } s + t < \frac{\alpha}{\beta}, \\ -cs & \text{if } s + t \geqslant \frac{\alpha}{\beta}, \end{cases}$$

$$\pi_2(s, t) = pt - c_2(t) = \begin{cases} (\alpha - \beta(s+t))t - ct & \text{if } s + t < \frac{\alpha}{\beta}, \\ -ct & \text{if } s + t \geqslant \frac{\alpha}{\beta}. \end{cases}$$

To calculate Player 2's best response function, we must maximize $\pi_2(s, t)$, s fixed, on the interval $0 \leqslant t < \infty$. This was done in Section 1.11; the answer is

$$t = b_2(s) = \begin{cases} \dfrac{\alpha - c}{2\beta} - \dfrac{1}{2}s & \text{if } s < \dfrac{\alpha - c}{\beta}, \\ 0 & \text{if } s \geqslant \dfrac{\alpha - c}{\beta}. \end{cases}$$

From the symmetry of the problem, Player 1's best response function is

$$s = b_1(t) = \begin{cases} \dfrac{\alpha - c}{2\beta} - \dfrac{1}{2}t & \text{if } t < \dfrac{\alpha - c}{\beta}, \\ 0 & \text{if } t \geqslant \dfrac{\alpha - c}{\beta}. \end{cases}$$

See Figure 3.1. Notice that to make the figure analogous to the payoff matrix in Section 3.9, which can be interpreted as the graph of a best response correspondence, we have made the s-axis, which represents Player 1's strategies, the vertical axis.

There is a Nash equilibrium where the two best response curves intersect. From the figure, we see that to find this point, we must solve simultaneously

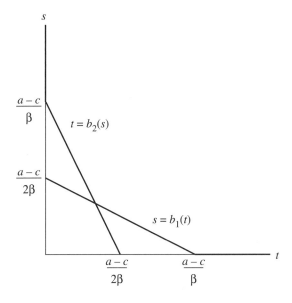

Figure 3.1. Best response functions in Cournot's model of duopoly.

the two equations

$$t = \frac{\alpha - c}{2\beta} - \frac{1}{2}s, \qquad s = \frac{\alpha - c}{2\beta} - \frac{1}{2}t.$$

We find that $s = t = (\alpha - c)/3\beta$.

3.12 Problems

3.12.1 Price Competition 1. Bernie and Mannie sell tablet computers. Both have unit costs of 100. They compete on price: the low-price seller gets all the customers. If they charge the same price, they split the customers. Explain why the only Nash equilibrium is for both to charge 100, splitting the market but making no profit. Suggestion: First set up the game by giving players, strategies, and payoffs. The players are Bernie and Mannie. Bernie's strategy is his price x and Mannie's strategy is his price y. Let's allow x and y to be any nonnegative real numbers. Let the the number of customers be $m > 0$. Each player's payoff is his profit, which is his fraction of the customers times m times profit per customer. Which strategy profiles (x, y) are Nash equilibria? For example, if $100 < x < y$, is (x, y) a Nash equilibrium? (Answer: no. Bernie gets all the customers and makes a profit of $x - 100 > 0$ on every tablet sold. Mannie gets nothing. Bernie could improve his payoff by increasing his price but keeping it less than y, or Mannie could improve his payoff by decreasing his price to a number between 100 and x.) Try to consider all possibilities in an organized way.

3.12.2 Price Competition 2.   Same problem as the previous one, except we add one more strategy for each player: charge 200, but advertise that if the tablet is available cheaper at the other store, the customer can have it for free. Show that now there are exactly two Nash equilibria: the previous one, and a new one in which both players use the new strategy.

3.12.3 Two Stores on Main Street.   Pleasantville has one street, Main Street. The residents of the town are uniformly distributed along Main Street between one end and the other.

Two companies are considering opening stores on Main Street. Each store must choose a location along the street between one end (0) and the other end (1). Each store will attract the fraction of the population that is closer to it than to the other store. If both stores locate at the same point (this is allowed), each will attract half the population. The payoff to each company is the fraction of the population that it attracts.

Assume that the two companies choose their locations simultaneously. The first company's strategy set is the set of real numbers x between 0 and 1, and the second company's strategy set is the set of real numbers y between 0 and 1. If $x = y$, each company's payoff is $\frac{1}{2}$. If $x < y$, company 1's payoff is $\frac{1}{2}(x + y)$, and company 2's payoff is $1 - \frac{1}{2}(x + y)$. If $x > y$, company 1's payoff is $1 - \frac{1}{2}(x + y)$, and company 2's payoff is $\frac{1}{2}(x + y)$.

Show that there is exactly one Nash equilibrium, $x = y = \frac{1}{2}$.

Suggestion to help you get started: If $x < y$, then company 1 can improve its payoff by changing to a new strategy x' that is between x and y.

3.12.4 Three Stores on Main Street.   Same problem as the previous one, but now there are three companies. Each attracts the fraction of the population that is closest to it. If two stores occupy the same location, they split the fraction of the population that is closer to them than to the other store. If all three stores occupy the same location, each attracts a third of the population. Show that there is no Nash equilibrium.

3.12.5 The Spoils of War.   Two countries each have one unit of wealth. Each chooses a fraction of its wealth to devote to fighting the other. The country that devotes a larger fraction of its wealth to fighting wins the fight. Its payoff is the remaining wealth of both countries. The losing country's payoff is zero. If both countries devote the same fraction of their wealth to fighting, the result is a tie. In this case, each country's payoff is its remaining wealth.

We consider this situation as a two-player game. The first country's strategy is a real number s, $0 \leqslant s \leqslant 1$, that represents the fraction of its wealth it will devote to fighting. Similarly, the second country's strategy is a real number t, $0 \leqslant t \leqslant 1$, that represents the fraction of its wealth it will devote to fighting. We assume the two countries choose their strategies simultaneously.

The payoffs are

- If $s < t$, $\pi_1(s,t) = 0$ and $\pi_2(s,t) = 2 - (s + t)$.
- If $s > t$, $\pi_1(s,t) = 2 - (s + t)$ and $\pi_2(s,t) = 0$.
- If $s = t$, $\pi_1(s,t) = 1 - s$ and $\pi_2(s,t) = 1 - t$. Of course, $1 - s = 1 - t$.

(1) Find all Nash equilibria with $s < t$. You may need to consider separately the case $t = 1$.
(2) Find all Nash equilibria with $s = t$.

On both parts, for each strategy profile (s,t), you should explain why it is or is not a Nash equilibrium.

3.12.6 Tit for Tat 1.   There are two toy stores in town, Al's and Bob's. If both charge high prices, both make $5K per week. If both charge low prices, both make $3K per week. If one charges high prices and one charges low prices, the one that charges high prices makes nothing, and the one that charges low prices makes $6K per week.

At the start of each week, both stores independently set their prices for the week.

Consider three possible strategies for each store:

- h: Always charge high prices.
- l: Always charge low prices.
- t: Tit for tat. Charge high prices the first week. The next week, do whatever the other store did the previous week.

The following matrix shows the payoffs if each store follows its strategy for two weeks:

		Bob		
		h	**l**	**t**
	h	(10,10)	(0,12)	(10,10)
Al	**l**	(12,0)	(6,6)	(9,3)
	t	(10,10)	(3,9)	(10,10)

(1) Suppose the t strategy were not available to either player. Explain why the remaining 2×2 game would be a prisoner's dilemma.
(2) Explain the $(9,3)$ payoffs in the second row of the matrix.
(3) Which of Al's strategies are strictly dominated?
(4) Which of Al's strategies are weakly dominated?

(5) Try to use iterated elimination of weakly dominated strategies to find a Nash equilibrium. How far do you get?

(6) Use best response to find all Nash equilibria.

3.12.7 Battle of the Sexes with Money Burning.

When there are several Nash equilibria, only one of which can be found using iterated elimination of weakly dominated strategies, is that Nash equilibrium in some sense the most reasonable one? Here is an example to think about. Consider the Battle of the Sexes game in Subsection 3.2.4, but change the payoff when Alice and Bob go to the same event to 3 for the one who prefers that event:

		Bob	
		concert	wrestling
Alice	concert	$(3,1)$	$(0,0)$
	wrestling	$(0,0)$	$(1,3)$

Now suppose that Alice has the option of burning some money before she leaves for the evening. If Alice does this, Bob will know, because of the smoke in the sky. If Alice burns some money, her payoff is reduced by 1 no matter what happens afterward.

Alice now has four strategies:

- bc: Burn money, then go to the concert.
- bw: Burn money, then go to the wrestling match.
- dc: Don't burn money, then go to the concert.
- dw: Don't burn money, then go to the wrestling match.

Bob also has four strategies: cc, cw, wc, and ww. The first letter indicates where Bob will go if he's sees the smoke; the second shows where he will go if he does not.

The payoff matrix is

		Bob			
		cc	cw	wc	ww
	bc	$(2,1)$	$(2,1)$	$(-1,0)$	$(-1,0)$
Alice	bw	$(-1,0)$	$(-1,0)$	$(0,3)$	$(0,3)$
	dc	$(3,1)$	$(0,0)$	$(3,1)$	$(0,0)$
	dw	$(0,0)$	$(1,3)$	$(0,0)$	$(1,3)$

(1) Use iterated elimination of weakly dominated strategies to find a Nash equilibrium.

(2) Use best response to find all Nash equilibria. (There are four.)

In this game, iterated elimination of weakly dominated strategies leads to Alice not burning money, then both go to the concert. Thus Alice gets her preferred outcome just by having available the option to burn some money, which is a ridiculous thing to do, and in fact she does not do it.

3.12.8 Should You Compromise? Which is better, to insist on doing what you want, or to compromise? Here is a simple model. Two friends, Players 1 and 2, must choose among activities a_1, a_2, and a_3. Player 1 prefers a_1, is neutral about a_2, and dislikes a_3. Player 2 is the reverse. Each player independently chooses an activity and gets a payoff of $v > 0$ if she picks the activity she prefers, 0 if she picks activity a_2, and $-v$ if she picks the activity she dislikes. In addition, each player incurs a cost $c > 0$ if both players choose the same activity. However, if a player is the only one to choose her activity, she incurs a higher cost rc, with $r > 1$. The factor r includes both the greater financial cost that often obtains when you do something alone, and the subjective cost of not having your friend with you.

The payoff matrix is therefore

		Player 2		
		a_1	a_2	a_3
	a_1	$(v-c, -v-c)$	$(v-rc, -rc)$	$(v-rc, v-rc)$
Player 1	a_2	$(-rc, -v-rc)$	$(-c, -c)$	$(-rc, v-rc)$
	a_3	$(-v-rc, -v-rc)$	$(-v-rc, -rc)$	$(-v-c, v-c)$

(1) Suppose $(r-1)c < v$. Find the Nash equilibria.
(2) Suppose $\frac{1}{2}(r-1)c < v < (r-1)c$. Find the Nash equilibria. (There are two. Which is best for each player?)
(3) Suppose $v < \frac{1}{2}(r-1)c$. Find the Nash equilibria. (There are three. For each player, which is best and which is second best?)
(4) Interpret the results.

3.12.9 The Twin Daughters. A mother has twin daughters whose birthday is approaching. She tells each to ask for a whole number of dollars between one and 100 (inclusive). If the total is 101 or less, each will get what she asks for. If the total exceeds 101, the daughters get nothing. Show that this game has exactly 100 Nash equilibria.

3.12.10 Avoiding Voters' Wrath. The three members of a city council are voting on whether to give themselves a raise. The raise passes if at least two members vote for it. The value of the raise to each member is $v > 0$, but each member who votes for it incurs a cost c, $0 < c < v$, in voter resentment.

The members of the city council are Alice, Bob, and Carol. Alice votes first; then Bob, knowing Alice's vote, votes; then Carol, knowing Alice's and Bob's votes, votes.

(1) Draw the game tree. (There are eight terminal vertices.)
(2) Use backward induction to find a Nash equilibrium.
(3) Show that there is another Nash equilibrium in which Carol's strategy is to vote no, whatever Alice and Bob do. Suggestion: There are too many strategies for you to use the normal form of the game: Alice has two strategies, Bob has four, and Carol has sixteen. Instead, assume that Carol uses the suggested strategy. Draw a 2×4 matrix that shows the payoffs to Alice and Bob from each of their possible strategy choices, assuming that Carol uses the suggested strategy. Find strategies for Alice and Bob that make a Nash equilibrium for them, assuming that Carol uses the suggested strategy. Once you've found these strategies, argue that Carol also cannot improve her own payoff by changing to one of her other strategies.

3.12.11 Grocery Store and Gas Station 2. Consider the first version of Grocery Store and Gas Station, Problem 1.14.6. Suppose the Grocery Store and the Gas Station choose their prices simultaneously. Find the Nash equilibrium.

3.12.12 Pollution, Health, and International Relations. Countries 1 and 2 both have coal-fired power plants, which pollute the air. If Country i burns x_i units of coal per year, its national income will increase by $3x_i$ billion dollars per year. One-fourth of the pollution produced in Country i crosses the border to the other country. Thus the pollution level in Country 1 will be $p_1 = \frac{3}{4}x_1 + \frac{1}{4}x_2$, and the pollution level in Country 2 will be $p_2 = \frac{1}{4}x_1 + \frac{3}{4}x_2$. Annual health costs to Country i of a pollution level p_i are p_i^2 billion dollars. Hence if Countries 1 and 2 burn x_1 and x_2 units of coal per year, respectively, their net gains to national income, in billions of dollars, are respectively,

$$\pi_1(x_1, x_2) = 3x_1 - \left(\tfrac{3}{4}x_1 + \tfrac{1}{4}x_2\right)^2,$$
$$\pi_2(x_1, x_2) = 3x_2 - \left(\tfrac{1}{4}x_1 + \tfrac{3}{4}x_2\right)^2.$$

We regard this as a two-player game. The countries choose x_1 and x_2; the payoffs are π_1 and π_2.

(1) Suppose the two countries simultaneously choose the quantities of coal to burn. Find the Nash equilibrium.
(2) Suppose Country 1 chooses x_1 first, then Country 2 observes x_1 and chooses x_2. Use backward induction to find Country 1's best choice.

3.12.13 Tobacco Market, continued. For the Tobacco Market example of Section 3.5, show that there are no Nash equilibria in which $q_1 + q_2 + q_3 <$ 975,000, exactly one q_i equals 600,000, and the other q_is are strictly between 0 and 600,000. Suggestion: Suppose it is q_1 that equals 600,000. Show that $(\partial \pi_1 / \partial q_1)(600,000, q_2, q_3) < 0$. This means that Farmer 1 can increase his payoff by reducing the quantity he brings a little.

3.12.14 Braess's Paradox. When a road network is crowded and travel times are slow, is it possible that adding another road to the network will make travel times even slower?

Consider the road network in Figure 3.2. Four thousand drivers want to go from S (start) to F (finish). Road e has not been built, so they can follow route ab or route cd. Roads b and c are modern, fast roads but roundabout. Travel time on each is 45 minutes at any reasonable traffic level. Roads a and d are short but easily overcrowded. If m drivers choose road a, travel time on it is $\frac{m}{100}$ minutes. Thus if all 4,000 drivers were to choose road a, travel time on it would be 40 minutes. Road d is the same.

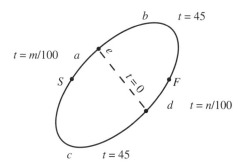

Figure 3.2. Roads and travel times from S to F.

We regard this situation as a game with 4,000 players, the drivers. The players simultaneously choose whether to follow route ab or route cd. If m drivers choose ab and n choose cd, the first m have travel time $\frac{m}{100} + 45$, and the second n have travel time $\frac{n}{100} + 45$. Of course, $m + n = 4000$. The payoff to each driver is minus her travel time. (Thus shorter travel times give higher payoffs.)

(1) Explain: there is a Nash equilibrium if 2000 drivers choose each route; if different numbers of drivers choose each route, it is not a Nash equilibrium. Notice that total travel time for each driver at a Nash equilibrium is $\frac{2000}{100} + 45 = 65$ minutes.

(2) Suppose road e is built as a modern road with travel time 0 at any reasonable traffic level. (This is for simplicity. Any travel time under 5 minutes gives a similar result.) Now two more routes are available: aed and ceb. Explain: the only Nash equilibrium is for every driver to choose aed. Unfortunately, total travel time for each driver is now 80 minutes.

For more on Braess's Paradox, see the Wikipedia page http://en.wikipedia .org/wiki/Braess's_paradox.

3.12.15 Sacred Places. Adam Smith compared established religions to monopolies, and religious freedom to competitive free enterprise. However, a third form of religious organization has occasionally existed in human history, in which religious power was vested in sacred places outside the domain of any ruler. Examples include the Oracle of Delphi during the period of independent Greek city-states, and the shrine at Shiloh at the time of the Book of Judges, when the tribes of Israel were independent. How do we explain the power of an entity such as the Oracle at Delphi, which the rulers of Greek city-states called on to resolve the most difficult disputes? Its decisions were respected even though it had no power to enforce them.

To model this situation, let us assume that there are $N \geqslant 2$ independent rulers of small domains, each with a sacred site in his capital city, and $M \geqslant 1$ independent sacred sites outside the domain of any ruler. Each ruler can publicly choose a sacred site at which to worship (his own, another ruler's, or an independent site), and exert efforts to get his citizens to share his allegiance. Alternatively, a ruler may choose not to publicly select a sacred site and to let his citizens do whatever they want.

Payoffs and costs are as follows:

- Each ruler who publicly chooses a sacred site and exerts efforts to get his citizens to do the same incurs a cost of $C > 0$.
- Each ruler who chooses the site in his own capital receives a payoff of $I > 0$ (I stands for "inside").
- Each ruler who chooses a site outside his own capital receives a payoff of $O > 0$ (O stands for "outside").
- Each ruler who chooses a site outside his own capital pays a "tax" $t > 0$. (This could represent, for example, gifts to the local temple, or spending by the visiting ruler.) If the site he chooses is in another ruler's capital, this tax is income to the other ruler.
- Each ruler receives a payoff in prestige that depends on the popularity of the site he chooses. The prestige he receives equals a number $l > 0$ times the number of outside rulers who worship at that site.

Therefore:

- If a ruler chooses the site in his own city, as do n other rulers, his payoff is $I - C + tn + ln$.
- If n other rulers choose the site in a ruler's city, but that ruler chooses a site elsewhere and is one of a total of m outside rulers to choose that site, the ruler's payoff is $tn + O - C - t + lm$.

We assume that $I > O - t + l$; that is, it is better to be the only ruler choosing the site in your own city than to be the only outside ruler choosing another site.

(1) Show that it is a Nash equilibrium for every ruler to choose the site in his own city.

(2) Show that if N is sufficiently large, it is a Nash equilibrium for every ruler to choose the site in one ruler's city. (To show this, consider, the possibility that an outside ruler switches to his own city. His payoff changes from $O - C - t + l(N - 1)$ to $I - C$. Is this an improvement? You could also consider the possibility that an outside ruler switches to another city, and the possibility that the inside ruler switches to another city, but these are worse than the previous possibility, so you can ignore them.) A similar argument shows that if N is sufficiently large, it is a Nash equilibrium for every ruler to choose one independent site.

(3) Show that if C is large enough, it is a Nash equilibrium for every ruler to let his citizens do as they please.

(4) Suppose N is even. Is it a Nash equilibrium for half the rulers to choose one independent site and half to choose another? Explain.

Chapter 4

Games in extensive form with incomplete information

In this chapter we return to games in extensive form, the subject of Chapter 1, but this time the players' knowledge of the game is incomplete. This may be because there are events in the game that cannot be predicted, such as the deal of a card or the outcome of a battle. Or it may be because there is an important fact that a player does not know. The notion of probability comes into play. We have waited until now to discuss these games because they cannot always be treated by backward induction. It may be necessary to convert the game into normal form and analyze that.

4.1 Utility functions and lotteries

In this section we introduce an idea needed in this chapter, the expected value of a lottery. We will illustrate lotteries in the next section by considering why people buy insurance. The explanation is related to utility functions and their commonly assumed properties, so we discuss utility functions first.

4.1.1 Utility functions.
A salary increase from $20,000 to $30,000 and a salary increase from $220,000 to $230,000 are not equivalent in their effect on your happiness. This is true even if you don't have to pay taxes!

Let s be your salary and $u(s)$ the *utility* of your salary to you. Two commonly assumed properties of $u(s)$ are:

(1) $u'(s) > 0$ for all s (strictly increasing utility function). In other words, more is better! If you have a strictly increasing utility function, either of those salary jumps will make you happier.

(2) $u''(s) < 0$ (strictly concave utility function). In other words, $u'(s)$ decreases as s increases. If your utility function is strictly concave in addition to being strictly increasing, the first salary jump, from $20,000 to $30,000, will increase your happiness more than the second one, from $220,000 to $230,000.

4.1.2 Lotteries. A *lottery* has n possible outcomes. The outcome depends on chance. The ith outcome occurs with probability p_i and yields a payoff x_i, which is a real number. We have all $p_i \geq 0$ and $p_1 + \cdots + p_n = 1$. The *expected value* of the lottery is

$$E[x] = p_1 x_1 + \cdots + p_n x_n.$$

If the recipient of the payoff has a utility function $u(x)$, its expected value is

$$E[u(x)] = p_1 u(x_1) + \cdots + p_n u(x_n).$$

The *expected utility principle* states that the lottery with the higher expected utility is preferred. For some discussion of the conditions under which this principle is true, see the Wikipedia page http://en.wikipedia.org/wiki/Expected_utility_hypothesis.

4.2 Buying Fire Insurance

This example is a decision problem, not a game.

You have a warehouse worth $1.2 million. The probability of fire in any given year is 5%. Fire insurance costs $100,000 per year. Should you buy it? To answer the question, we compare the two lotteries shown in Figure 4.1. Without fire insurance the expected payoff is

$$E[x] = .05 \times -1.2M + .95 \times 0 = -60K.$$

With fire insurance the expected payoff is

$$E[x] = .05 \times -100K + .95 \times -100K = -100K.$$

Don't buy the insurance.

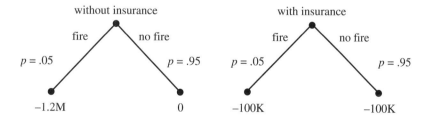

Figure 4.1. Should you buy the insurance?

However, people typically have concave utility functions, as we saw in Section 4.1. Suppose your utility function is $u(x) = \ln(1.3M + x)$. (This function is just $\ln x$ shifted 1.3M to the left, so it is continuous on the interval

$-1.3\text{M} < x < \infty$.) Now, without fire insurance the expected utility payoff is

$$E[u(x)] = .05\ln(.1\text{M}) + .95\ln(1.3\text{M}) = 13.95.$$

With fire insurance, the expected utility payoff is

$$E[u(x)] = .05\ln(1.2\text{M}) + .95\ln(1.2\text{M}) = 14.00.$$

Buy the insurance.

With a strictly increasing, strictly concave utility function, as your gain increases, your utility rises more and more slowly. By the same token, as your gain decreases, or your loss increases, your utility falls more and more quickly! In other words, big gains are good, but big losses are terrible. This is a reason that people buy insurance against the possibility of big losses.

4.3 Games in extensive form with incomplete information

To treat games in which players have incomplete information, we add two ingredients to our allowed models of games in extensive form:

- Certain nodes may be assigned not to a player but to Nature. Nature's moves are chosen by chance. Therefore, if c is a node assigned to Nature, each move that starts at c will be assigned a probability $0 \leqslant p \leqslant 1$, and these probabilities will sum to 1.
- The nodes assigned to a player may be partitioned into *information sets*. If several of a player's nodes are in the same information set, then the player does not know which of these nodes represents the true state of the game. The sets of available moves at the different nodes of an information set must be identical.

Each path in the game tree is assigned a probability: probability 1 if the path does not include any of Nature's moves, and the product of the probabilities of Nature's moves along the path if it does.

A player's strategy is required to assign the same move to every node of an information set.

A strategy profile σ determines a collection of complete paths $C(\sigma)$ through the game tree. The collection $C(\sigma)$ may include more than one complete path. To determine the expected payoff to a player of a strategy profile, one sums the payoffs of the complete paths in the collection $C(\sigma)$, each multiplied by the probability of the path.

A node assigned to Nature may represent a random event, such as the deal of a card, or it may represent a point at which a player does not know what the situation is but can assign probabilities to the different possibilities. In either case, one player may have information about Nature's moves that other players lack. For example, in a card game, Nature decides the hand you are dealt. You know it, but other players do not.

Backward induction often does not work in games in extensive form with incomplete information. The difficulty comes when you must decide on a move at an information set that includes more than one node. Usually the payoffs for the available moves depend on which node you are truly at, but you don't know that.

If the only nodes that precede the information set in question are Nature's, then you can calculate the probability that you are at each node and use expected value to make a choice. In our treatment of the Cuban Missile Crisis in Section 4.6, we encounter information sets of this type. In contrast, if, preceding the information set in question, there are nodes at which other players made a choice, then you do not know the probability that you are at each node. Games with information sets of this type can be treated by converting them to games in normal form. The game Buying a Used Car in the next section is an example.

4.4 Buying a Used Car

A customer is interested in a used car. She asks the salesman if it is worth the price. The salesman wants to sell the car. He also wants a reputation for telling the truth. How does the salesman respond? And should the customer believe his response?

We assume that for cars of the type being sold, the probability the car is worth the price (i.e., is a good car) is p, so the probability it is not worth the price (i.e., is a bad car) is $1 - p$. When we complete our analysis we will consider whether the customer or salesman needs to know these probabilities. However, the salesman knows whether this particular car is good or bad. The customer does not.

The payoffs are:

- The salesman gets 2 points if he sells the car and 1 point if he tells the truth.
- The customer gets 1 point if she correctly figures out whether the car is good or bad.

We model this situation as a game in extensive form with incomplete information. Nature moves first and decides whether the car is good or bad. Then the salesman tells the customer whether it is good or bad. Then the customer decides whether it is good or bad, and on that basis decides whether to buy

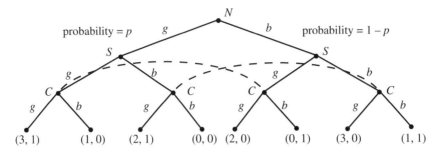

Figure 4.2. Buying a used car. The salesman's payoffs are given first.

it. See Figure 4.2. Notice that the customer's moves are divided into two information sets, reflecting the fact that when the salesman says the car is good or bad, the customer does not know whether that is true.

This game is an example of a game in extensive form with incomplete information in which backward induction cannot be used. For example, what should the customer do if the salesman says the car is good? The corresponding information set contains two nodes: at one of them, the car is actually good, but at the other it is bad. The customer's best move depends on which is the case, but she doesn't know which is the case. More importantly, she doesn't know the probabilities, because there are nodes preceding the nodes in this information set at which the salesman made a choice of what to say. If the salesperson speaks randomly, then when he says the car is good, it really is good with probability p. However, there is no reason to believe that the salesman speaks randomly.

We therefore analyze this game by converting it to a game in normal form. The salesman has four strategies:

- *gg*: If the car is good, say it is good; if the car is bad, say it is good. (Always say the car is good.)
- *gb*: If the car is good, say it is good; if the car is bad, say it is bad. (Always tell the truth.)
- *bg*: If the car is good, say it is bad; if the car is bad, say it is good. (Always lie.)
- *bb*: If the car is good, say it is bad; if the car is bad, say it is bad. (Always say the car is bad.)

The customer must make the same move at two nodes that are in the same information set. Thus she only has four strategies:

- *gg*: If the salesman says the car is good, believe it is good; if the salesman says the car is bad, believe it is good. (Always believe the car is good.)

- gb: If the salesman says the car is good, believe it is good; if the salesman says the car is bad, believe it is bad. (Always believe the salesman.)
- bg: If the salesman says the car is good, believe it is bad; if the salesman says the car is bad, believe it is good. (Never believe the salesman.)
- bb: If the salesman says the car is good, believe it is bad; if the salesman says the car is bad, believe it is bad. (Always believe the car is bad.)

Consider the salesman to be Player 1 and the customer to be Player 2.

In this game, each strategy profile is associated with *two* paths through the game tree. For example, consider the strategy profile (gg, bg): the salesman always says the car is good; and the customer never believes the salesman. This profile is associated with the two paths ggb and bgb:

- ggb: Nature decides the car is good, the salesman says the car is good, the customer does not believe him and decides it is bad (and hence does not buy the car): Payoffs: 1 to the salesman for telling the truth, 0 to the customer for miscalculating.
- bgb: Nature decides the car is bad, the salesman says the car is good, the customer does not believe him and decides it is bad (and hence does not buy the car): Payoffs: 0 to the salesman (he lied and still didn't sell the car), 1 to the customer (for correctly deciding the car was bad).

The first path through the tree is assigned probability p (the probability of a good car), the second is assigned probability $1 - p$ (the probability of a bad car). Thus the ordered pair of payoffs assigned to the strategy profile (gg, bg) is $p(1, 0) + (1 - p)(0, 1) = (p, 1 - p)$.

In this game, a good way to derive the payoff matrix is to separately write down the payoff matrix when Nature chooses a good car and when Nature chooses a bad car. The payoff matrix for the game is then p times the first matrix plus $1 - p$ times the second.

If the car is good, the payoffs are

		Customer			
		gg	gb	bg	bb
	gg	$(3,1)$	$(3,1)$	$(1,0)$	$(1,0)$
Salesman	gb	$(3,1)$	$(3,1)$	$(1,0)$	$(1,0)$
	bg	$(2,1)$	$(0,0)$	$(2,1)$	$(0,0)$
	bb	$(2,1)$	$(0,0)$	$(2,1)$	$(0,0)$

If the car is bad, the payoffs are

		Customer			
		gg	gb	bg	bb
Salesman	gg	$(2,0)$	$(2,0)$	$(0,1)$	$(0,1)$
	gb	$(3,0)$	$(1,1)$	$(3,0)$	$(1,1)$
	bg	$(2,0)$	$(2,0)$	$(0,1)$	$(0,1)$
	bb	$(3,0)$	$(1,1)$	$(3,0)$	$(1,1)$

The payoff matrix for the game is p times the first matrix plus $1 - p$ times the second:

		Customer			
		gg	gb	bg	bb
Salesman	gg	$(2+p,p)$	$(2+p,p)$	$(p,1-p)$	$(p,1-p)$
	gb	$(3,p)$	$(1+2p,1)$	$(3-2p,0)$	$(1,1-p)$
	bg	$(2,p)$	$(2-2p,0)$	$(2p,1)$	$(0,1-p)$
	bb	$(3-p,p)$	$(1-p,1-p)$	$(3-p,p)$	$(1-p,1-p)$

Let us assume $p > \frac{1}{2}$, so the car is usually good. We look for Nash equilibria using best response, so we draw a box around the highest first entry in each column and the highest second entry in each row (see Section 3.9). The assumption $p > \frac{1}{2}$ is needed to find the Customer's best response to the Salesman's strategies gg and bb.

		Customer			
		gg	gb	bg	bb
Salesman	gg	$(2+p,\boxed{p})$	$(\boxed{2+p},\boxed{p})$	$(p,1-p)$	$(p,1-p)$
	gb	$(\boxed{3},p)$	$(1+2p,\boxed{1})$	$(3-2p,0)$	$(\boxed{1},1-p)$
	bg	$(2,p)$	$(2-2p,0)$	$(2p,\boxed{1})$	$(0,1-p)$
	bb	$(3-p,\boxed{p})$	$(1-p,1-p)$	$(\boxed{3-p},\boxed{p})$	$(1-p,1-p)$

There are two Nash equilibria, (gg,gb) and (bb,bg). At the first equilibrium, the salesman always says the car is good, and the customer always believes the salesman. At the second, the salesman always says the car is bad, and the customer always assumes the salesman is lying. The two Nash equilibria give the same payoff to the customer, but the first gives a better payoff to the salesman.

Now that we have completed the analysis, we see that to find the Nash equilibrium of the game, the salesman and customer only need to know that $p > \frac{1}{2}$; they do not need any more detailed knowledge.

4.5   The Travails of Boss Gorilla 1

Boss Gorilla is boss of a Gorilla Band. Other gorillas are out there who might challenge him. When a visiting gorilla appears, it can do one of two things:

- Challenge Boss Gorilla.
- Leave.

If a visiting gorilla challenges Boss Gorilla, Boss Gorilla has two choices:

- Acquiesce. In this case the visiting gorilla joins the Gorilla Band and becomes co-boss with Boss Gorilla.
- Fight.

There are two types of visiting gorillas: Tough and Weak. The Tough ones will win a fight with Boss Gorilla. The Weak ones will lose.

Boss Gorilla believes that the probability that a visiting gorilla is Tough is p, with $0 < p < 1$. The probability that he is Weak is $1 - p$. A visiting gorilla knows which type he is, but Boss Gorilla does not.

We view this as a game with three players: Player 1, Tough Gorilla (T); Player 2, Weak Gorilla (W); and Player 3, Boss Gorilla (B). Figure 4.3 illustrates the situation and gives the payoffs.

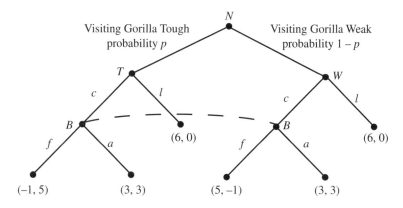

Figure 4.3. T is Tough Visiting Gorilla, W is Weak Visiting Gorilla, and B is Boss Gorilla. Boss Gorilla's payoffs are given first, the visiting gorilla's payoffs are given second. If a visiting gorilla leaves, payoffs are 0 to him and 6 to Boss Gorilla. This is the value of being boss of the Gorilla Band. Therefore, if a visiting gorilla challenges and Boss Gorilla acquiesces, payoffs are 3 to each. If the gorillas fight, payoffs are −1 to the loser (for injuries sustained) and 5 to the winner (6 for getting to be boss of the Gorilla Band, minus 1 for injuries sustained). The dashed line connects nodes in the same information set: if a visiting gorilla challenges, Boss Gorilla does not know whether the visitor is Tough or Weak.

This game is similar to Buying a Used Car in that backward induction cannot be used. If the visiting gorilla challenges, Boss Gorilla does not know whether the visitor is Tough or Weak, or even what the probabilities are, since choices were made previously. Hence we should convert to a game in normal form. Tough Gorilla has two strategies: challenge (*c*) or leave (*l*). Weak Gorilla has the same two strategies. Boss Gorilla has two strategies: fight if challenged (*f*) or acquiesce if challenged (*a*).

Although this game is most naturally conceived as one with three players, it can be successfully analyzed by viewing it as a game with two players, Visiting Gorilla and Boss Gorilla. In this way of looking at the game, Nature first decides the type of Visiting Gorilla, Tough or Weak, and then a two-player game is played between Visiting Gorilla and Boss Gorilla. The game tree for this two-player game is exactly that shown in Figure 4.3, except that the nodes labeled *T* (Tough Gorilla) and *W* (Weak Gorilla) should both be labeled *V* (Visiting Gorilla). Visiting Gorilla is assumed to know his own type, so he can use different strategies in the two cases.

Thus, when the two-player game is converted to normal form, Visiting Gorilla has four strategies:

- *cc*: always challenge.
- *cl*: challenge if Tough, leave if Weak.
- *lc*: leave if Tough, challenge if Weak.
- *ll*: always leave.

Boss Gorilla, however, does not know the type of Visiting Gorilla, so he just has two strategies: fight if challenged (*f*) or acquiesce if challenged (*a*).

Best responses in the three-player game are related to best responses in the two-player game. For example, if Boss Gorilla uses the strategy fight if challenged, in the three player-game the best response of Tough Gorilla is to challenge (*c*) and the best response of Weak Gorilla is to leave (*l*). Therefore in the two-player game the best response of Visiting Gorilla is the strategy *cl*.

Similarly, suppose in the three-player game, Tough Gorilla uses *c* and Visiting Gorilla uses *l*. Boss Gorilla's best response will also be his his best response to the strategy *cl* in the two-player game.

It follows that Nash equilibria of the two-player game correspond to Nash equilibria of the three-player game, so the simpler two-player game can be analyzed instead.

The payoff matrix for the two-player game can be calculated like that in Buying a Used Car (Section 4.4). If Visiting Gorilla is Tough, the payoffs are

		Visiting Gorilla			
		cc	*cl*	*lc*	*ll*
Boss Gorilla	*f*	$(-1, 5)$	$(-1, 5)$	$(6, 0)$	$(6, 0)$
	a	$(3, 3)$	$(3, 3)$	$(6, 0)$	$(6, 0)$

If Visiting Gorilla is Weak, the payoffs are

		Visiting Gorilla			
		cc	*cl*	*lc*	*ll*
Boss	*f*	(5, −1)	(6, 0)	(5, −1)	(6, 0)
Gorilla	*a*	(3, 3)	(6, 0)	(3, 3)	(6, 0)

The payoff matrix for the two-player game is p times the first matrix plus $1 - p$ times the second:

		Visiting Gorilla			
		cc	*cl*	*lc*	*ll*
Boss	*f*	$(5 - 6p, 6p - 1)$	$(6 - 7p, 5p)$	$(5 + p, p - 1)$	$(6, 0)$
Gorilla	*a*	$(3, 3)$	$(6 - 3p, 3p)$	$(3 + 3p, 3 - 3p)$	$(6, 0)$

Boss Gorilla's best responses to cl, lc, and ll are the same for any p with $0 < p < 1$. Visiting Gorilla's best responses to f and a are also the same for any p with $0 < p < 1$. These best responses are indicated in the following table.

		Visiting Gorilla			
		cc	*cl*	*lc*	*ll*
Boss	*f*	$(5 - 6p, 6p - 1)$	$(6 - 7p, \boxed{5p})$	$(\boxed{5 + p}, p - 1)$	$(\boxed{6}, 0)$
Gorilla	*a*	$(3, \boxed{3})$	$(\boxed{6 - 3p}, 3p)$	$(3 + 3p, 3 - 3p)$	$(\boxed{6}, 0)$

However, Boss Gorilla's best response to cc depends on the value of p: we have $3 > 5 - 6p$ if and only if $6p > 2$; that is, $p > \frac{1}{3}$. Hence for $p > \frac{1}{3}$, Boss Gorilla's best response to cc is a, so the two-player game has the Nash equilibrium (a, cc).

In the corresponding Nash equilibrium of the three-player game, both visiting gorillas challenge, and Boss Gorilla acquiesces. In other words, when the probability of a Tough Visiting Gorilla is high enough ($p > \frac{1}{3}$), Boss Gorilla cannot risk fighting; Weak Visiting Gorilla takes advantage of the situation by always challenging.

We treat the case $p < \frac{1}{3}$ in the next chapter (Problem 5.12.2).

We could have made the analysis a little easier by noticing early that Tough Visiting Gorilla should always challenge; challenge gives Tough Visiting Gorilla a payoff of at least 3, whereas leave gives Tough Visiting Gorilla a payoff of 0. Thus we could eliminate from the beginning Visiting Gorilla's two strategies lc and ll, which use leave when Visiting Gorilla is Tough. We would only have to deal with 2×2 matrices instead of 2×4 matrices.

Another way to explain why Visiting Gorilla's two strategies lc and ll can be eliminated is that each is strictly dominated: cc strictly dominates lc, and

cl strictly dominates ll. You can easily check this from the payoff matrix above, but it is more enlightening to realize the relation between the fact that these strategies are dominated and the fact that Tough Visiting Gorilla should always challenge.

Let's work out in detail why cc strictly dominates lc, using the fact that Tough Visiting Gorilla should always challenge:

- cc uses c and lc uses l when Visiting Gorilla is Strong, so in this case cc gives a better payoff to Visiting Gorilla than lc against either of Boss Gorilla's strategies. Against f, for example, cc gives 5 and lc gives 0.
- cc and lc both use the same choice c when Visiting Gorilla is Weak, so in this case cc and lc give Visiting Gorilla the same payoff against either of Boss Gorilla's strategies. Against f, for example, both give -1.
- Against one of Boss Gorilla's strategies, the total payoff to Visiting Gorilla from cc or lc is a weighted average of his payoffs when Strong and Weak. Since cc gives a better payoff than lc when Visiting Gorilla is Strong and the same payoff when Visiting Gorilla is Weak, the weighted average payoff from cc must be greater than the weighted average payoff from lc. Against f, for example, Visiting Gorilla's weighted average payoff from cc is $p \cdot 5 + (1-p) \cdot (-1)$, and his weighted average payoff from lc is $p \cdot 0 + (1-p) \cdot (-1)$. Since $5 > 0$ and $0 < p < 1$, the first is greater.

We conclude that Visiting Gorilla's strategy cc strictly dominates his strategy lc.

A similar argument explains why cl strictly dominates ll.

In Section 4.6 we encounter another situation in which a player is unsure of the type of another player but has beliefs about the probabilities. Again we model this situation by assuming a move by Nature that determines the type of the opponent, and we will assume that whatever the type, the opponent wants to maximize it's own payoffs.

4.6 Cuban Missile Crisis

You may want to compare our account of the Cuban Missile Crisis to the Wikipedia article http://en.wikipedia.org/wiki/Cuban_missile_crisis.

In late summer of 1962, the Soviet Union began to place about 40 nuclear-armed medium- and intermediate-range ballistic missiles in Cuba. These missiles could target most of the eastern United States. The missile sites were guarded by surface-to-air missiles. There were also bombers.

A U.S. spy plane discovered the missiles on October 14, 1962. In the view of the U.S. government, the missiles posed several dangers: (1) they were a direct military threat to the United States, and could perhaps be used to compel U.S.

withdrawal from contested territories, such as Berlin; (2) they promised to deter any possible U.S. attack against Cuba; and (3) their successful placement in Cuba would be seen by the world as a Soviet victory and a U.S. defeat.

President John F. Kennedy and his associates at first considered an air strike against the missiles sites. Military leaders argued for a massive air strike against airfields and other targets as well. The civilian leaders decided these ideas were too risky and settled on a naval blockade.

The blockade went into effect October 24. Several apparently civilian freighters were allowed through with minimal inspection. Other questionable Soviet ships were heading toward Cuba, however, and the president and his associates feared that a confrontation with them could get out of hand. The Soviet premier, Nikita Khrushchev, indicated he might be willing to remove the missiles from Cuba if the United States removed its own missiles from Turkey. The United States had been planning to remove these missiles anyway and to substitute missiles on nuclear submarines but did not want to appear to be giving in to pressure or to make the Turks feel that the United States would not protect them.

On October 26 the United States discovered that the Soviets had also installed tactical nuclear weapons in Cuba that could be used against invading troops. On October 27 a U.S. spy plane was shot down over Cuba, and Cuban antiaircraft defenses fired on other U.S. aircraft. The U.S. Air Force commander, General Curtis Lemay, sent U.S. nuclear-armed bombers toward the Soviet Union, past their normal turnaround points.

On October 28 the crisis suddenly ended. Khrushchev announced that the missiles and other nuclear weapons in Cuba would be dismantled and brought home. Negotiations over the next month resulted in the withdrawal of the Soviet bombers as well, and, in a semisecret agreement, the removal of U.S. missiles from Turkey.

Today more is known about Soviet intentions in the crisis than was known to the U.S. government at the time. On the Soviet side, as on the U.S. side, there was considerable division over what to do. Apparently Khrushchev made the decision on his own to install missiles in Cuba; some of his advisors thought it reckless. He thought Kennedy would accept the missiles as a *fait accompli*, and he planned to issue an ultimatum to resolve the Berlin issue, using the missiles as a threat. The Soviet Presidium apparently decided as early as October 22 that it would back down rather than allow the crisis to lead to war. Two years later it removed Khrushchev from power. One if its main charges against him was the disastrous Cuban adventure.

On both sides it was not certain that decisions taken by leaders would be carried out as they intended. On the U.S. side, civilian leaders proposed methods of carrying out the blockade, but the U.S. Navy mostly followed its standard procedures. General LeMay, the inspiration for General Jack D. Ripper in the 1963 movie *Dr. Strangelove*, acted on his own in sending bombers toward the Soviet Union. He regarded the end of the crisis as a U.S. defeat: "We lost! We ought to just go in there today and knock 'em off." On the Soviet side, the decision to shoot down a U.S. spy plane on October 27 was taken by the deputy to the Soviet general in charge while the general was away from his desk. The Cubans' decision to fire on U.S. aircraft was taken by the Cuban president Fidel Castro over objections from the Soviet ambassador.

We consider several models of the Cuban Missile Crisis, beginning at the point where the missiles were discovered. A very simple model captures the essence of what happened: the United States can either accept the missiles or threaten war to remove them; if war is threatened, the Soviets can either defy the United States, which would lead to war, or can back down and remove the missiles. Backward induction should tell us what the two parties will do.

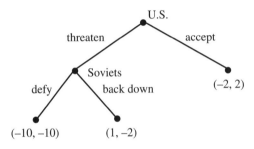

Figure 4.4. A simple model of the Cuban Missile Crisis. Payoffs to the United States are given first.

If the United States accepts the Soviet missiles in Cuba, we take the payoffs to be −2 to the United States and 2 to the Soviets. If the United States threatens war and the Soviets back down, we take the payoffs to be 1 to the United States and −2 to the Soviets. If the United States threatens war and the Soviets do not back down, we take the payoffs to be −10 to both the United States and the Soviets. See Figure 4.4.

We conclude that if the United States threatens war, the Soviets will back down. Using backward induction, the United States decides to threaten war rather than accept the Soviet missiles. The Soviets then back down rather than go to war.

This is in fact what happened. However, the Soviets could do this analysis, too, so why did they place the missiles in Cuba to begin with?

The payoffs in Figure 4.4 assume a rather reasonable Soviet leadership. If the Soviet leadership regarded backing down in the face of a U.S. threat as totally unacceptable, and was less fearful of nuclear war, we get a situation like that in Figure 4.5.

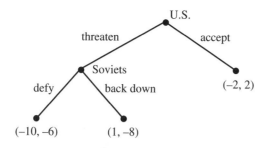

Figure 4.5. Cuban Missile Crisis with hardline Soviets.
Payoffs to the United States are given first.

In this case, if the United States threatens war, the Soviets will not back down. Using backward induction, the United States will decide to accept the missiles.

In fact the U.S. government was not sure whether the Soviets would turn out to be reasonable or hardline. The evidence was conflicting; for example, an accommodating letter from Khrushchev on October 24 was followed by the shooting down of a U.S. plane on October 26. In addition, the United States understood in general terms that the Soviet leaders were divided. Figure 4.6 shows the situation if the United States is unsure whether the Soviet leadership is reasonable, as in Figure 4.4, or hardline, as in Figure 4.5.

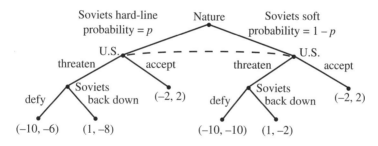

Figure 4.6. Cuban Missile Crisis with unknown Soviet leadership.
Payoffs to the United States are given first.

In the first move of the game, Nature decides, with certain probabilities, whether the Soviets are hardline or reasonable. The Soviets know which they are, but the United States does not. Therefore, when the United States makes its move, which is the first move in the game by a player, both its nodes are in the same information set; this is indicated by a dashed line in Figure 4.6. The United States must make the same move (threaten war or accept the missiles) at both of these nodes. However, if the United States threatens war, the Soviets, knowing who they are, will reply differently in the two cases.

This game differs from Buying a Used Car and Travails of Boss Gorilla in that the nodes toward the bottom of the tree are not in information sets containing more than one node. We can therefore at least start to analyze it by backward induction. In fact, the one information set that contains more than one move is preceded only by a move of Nature's. This kind of information set allows backward induction taking into account the probabilities.

We must look first at the two nodes where the Soviets move, following a U.S. threat of war. (These are the only nodes that are followed only by terminal vertices.) If the Soviets are hardline, they will choose to defy, with payoffs $(-10, -6)$. If the Soviets are reasonable, they will choose to back down, with payoffs $(1, -2)$.

Proceeding by backward induction, we look next at the two nodes where the United States moves, which are in the same information set. In this case, the probability that we are at each node in the information set is clear, so we can describe the payoffs of our choices using expected value.

If the United States threatens war, the payoffs are

$$p(-10, -6) + (1 - p)(1, -2) = (1 - 11p, -2 - 4p).$$

If the United States accepts the Soviet missiles, the payoffs are

$$p(-2, 2) + (1 - p)(-2, 2) = (-2, 2).$$

The United States will threaten war provided $1 - 11p > -2$, that is, provided $p < \frac{3}{11}$. If $p > \frac{3}{11}$, then the United States will accept the missiles.

President Kennedy apparently considered the probability of an unreasonable Soviet leadership to be somewhere between $\frac{1}{3}$ and $\frac{1}{2}$. Even $\frac{1}{3}$ is greater than $\frac{3}{11}$. Now the question is different from the one we posed after our first model: why didn't the United States accept the missiles?

In fact, the United States did not exactly threaten the Soviets with war if they did not remove their missiles. Instead it took actions, including a naval blockade, increased overflights of Cuba, and other military preparations, that increased the chance of war, even if war was not the U.S. intention. Both sides recognized that commanders on the scene might take actions the leadership did not intend, and events might spiral out of control. As we have seen, in the course of the crisis, dangerous decisions were in fact taken that leaders had trouble interpreting and bringing under control.

Brinkmanship (Wikipedia article: http://en.wikipedia.org/wiki/Brinkmanship) typically refers to the creation of a probabilistic danger to win a better outcome. On the one hand, it requires a great enough probability of disaster to persuade a reasonable opponent to concede rather than face the possibility that the dangerous situation will get out of hand. On the other hand, it requires a low probability of the nightmare outcome: the dangerous situation gets out of hand, *and* the opponent turns out to be a hardliner who will not back down. The term dates to the 1950s. It is generally believed that the United States successfully practiced brinkmanship in the Cuban Missile Crisis. Figure 4.7 is a brinkmanship model of the Cuban Missile Crisis.

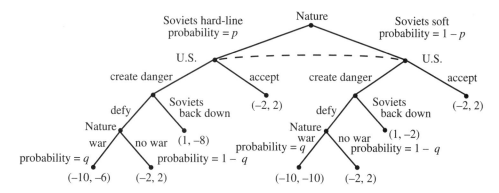

Figure 4.7. Cuban Missile Crisis with brinkmanship.
Payoffs to the United States are given first.

The change from Figure 4.6 is that, instead of threatening war, the United States now creates a dangerous situation by its military moves. The Soviets can defy or back down. If they defy, Nature decides whether there is war (with probability q) or no war (with probability $1 - q$). If there is war, the payoffs are what they were in Figure 4.6 if the Soviets defied the U.S. threat. If there is no war, then the missiles remain; we take the payoffs to be the same as those when the United States accepts the Soviet missiles to begin with.

The game in Figure 4.7 can be analyzed by backward induction. Nature's last move results in the following expected payoffs if the Soviets defy the United States:

- If Soviets are hardline: $q(-10, -6) + (1 - q)(-2, 2) = (-2 - 8q, 2 - 8q)$.
- If Soviets are reasonable: $q(-10, -10) + (1-q)(-2, 2) = (-2 - 8q, 2 - 12q)$.

Backing up one step, we find

- If the Soviets are hardline and the United States creates a dangerous situation, the Soviets will defy if $2 - 8q > -8$. Since this inequality is true for all q between 0 and 1, the Soviets will certainly defy.
- If the Soviets are reasonable and the United States creates a dangerous situation, the Soviets will defy if $2 - 12q > -2$, that is, if $q < \frac{1}{3}$. If $q > \frac{1}{3}$, the Soviets will back down.

Now we back up one more step and ask whether the United States should create a dangerous situation or accept the Soviet missiles. If the United States accepts the Soviet missiles, the payoffs are of course $p(-2,2) + (1 - p)(-2,2) = (-2,2)$. If the United States creates a dangerous situation, the payoffs depend on the probability of war q associated with the situation that the United States creates. The payoffs are

- If $q < \frac{1}{3}$: $p(-2 - 8q, 2 - 8q) + (1 - p)(-2 - 8q, 2 - 12q) = (-2 - 8q, 2 - 12q + 4pq)$.
- If $q > \frac{1}{3}$: $p(-2 - 8q, 2 - 8q) + (1 - p)(1, -2) = (1 - 3p - 8pq, -2 + 4p - 8pq)$.

Let's consider both cases.

- The payoff to the United States from creating a dangerous situation with $q < \frac{1}{3}$ is $-2 - 8q$. The United States is better off simply accepting with missiles, which yields a payoff of -2. The probability of war is too low to induce even reasonable Soviets to back down. Making such a threat increases the danger to the United States without any offsetting benefit.
- The payoff to the United States from creating a dangerous situation with $q > \frac{1}{3}$ is $1 - 3p - 8pq$. The United States benefits from creating such a situation if $1 - 3p - 8pq > -2$, that is, if

$$q < \frac{3(1 - p)}{8p}.$$

Therefore, if $\frac{1}{3} < 3(1 - p)/8p$, the United States can benefit by creating a dangerous situation in which the probability of war q is any number between $\frac{1}{3}$ and $3(1 - p)/8p$.

Figure 4.8 helps in interpreting this result. If $0 < p < \frac{3}{11}$, any q between $\frac{1}{3}$ and 1 gives the United States a better result than accepting the missiles. Of course, we already knew that for p in this range, a simple threat of war (equivalent to $q = 1$) would give the United States a better result than accepting the missiles. More interesting is the interval $\frac{3}{11} < p < \frac{9}{17}$, which includes the U.S. government's guess as to the true value of p. For each p in this interval there is a corresponding interval $\frac{1}{3} < q < 3(1 - p)/8p$ that gives the United States a better result than accepting the missiles.

4.7 Problems

4.7.1 Survivor Strategy. This problem is related to Section 4.2: you are asked to figure out which of two courses of action has the higher expected payoff.

At the end of season 1 of the television show *Survivor*, there were three contestants left on the island: Rudy, Kelly, and Rich. They were engaged in

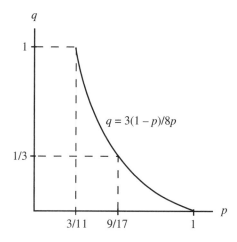

Figure 4.8. Where brinkmanship is helpful.

an "immunity challenge," in this case a stamina contest. Each contestant had to stand on an awkward support with one hand on a central pole. If the contestant's hand lost contact with the pole, even for an instant, the contestant was out. Once two contestants were out, the third contestant was the winner of the immunity challenge.

The winner of the immunity challenge would then choose one of the other two contestants to kick off the island.

Once there were only two contestants remaining on the island, a jury consisting of seven contestants who had recently been voted off the island would decide which of the two was the winner. The winner would get $1 million.

We pick up the story when the immunity contest has been going for $1\frac{1}{2}$ hours. Rudy, Kelly, and Rich are still touching the pole. Rich has been thinking about the following considerations (as he later explained to the camera):

- Rich and Kelly are strong young people. Rudy is much older. In addition, Kelly has become known for her stamina. Rich estimates that the probability of each winning the contest is Rich .45, Kelly .50, and Rudy .05. Rich further estimates that if he is the first contestant to lose touch with the pole, Kelly's probability of winning would be .9, and Rudy's would be .1.
- Rudy is much more popular with the jury than either Rich or Kelly. Rich figures that if Rudy is one of the last two contestants on the island, the jury is certain to pick Rudy as the winner.
- Rich and Kelly are equally popular with the jurors. However, if Rich or Kelly wins the immunity contest and kicks the popular Rudy off the island, some jurors might be made unhappy. Rich estimates that if he and Kelly are the last contestants on the island, but he has kicked off Rudy, there is a .4 chance the jury would pick him and a .6 chance it would pick Kelly. In

contrast, if he and Kelly are the last contestants on the island, and Kelly has kicked off Rudy, there is a .6 chance the jury would pick him and a .4 chance it would pick Kelly.

Rich is thinking about stepping away from the pole, thereby losing the immunity contest on purpose. Should he do it?

Make sure your instructor can follow your reasoning.

Figure 4.9 illustrates the situation.

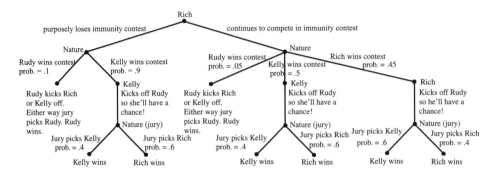

Figure 4.9. Survivor strategy.

(If you want to know what actually happened, you can certainly find the episode to watch!)

4.7.2 How to Control Crime.

This problem treats games in normal form like those in Chapter 3. However, the payoffs are expected payoffs, as introduced in this chapter.

In a town with no crime, a resident is considering committing a crime. If he does, the police will devote all their resources to investigating it. If the criminal is not caught, the crime will produce a benefit b to the criminal. If he is caught, he will suffer a punishment with value f. The probability of being caught is .1. Thus the criminal's expected gain from the crime is $.9b - .1f$, whereas his expected gain from not committing the crime is 0. If $f > 9b$, the resident will not commit the crime. In other words, if punishment is sufficiently high ($f > 9b$), then current police resources (which make the probability of getting caught .1) are sufficient to deter crime.

(1) Now suppose two residents, Al and Bob, are considering committing crimes that if successful will produce a benefit b to the criminal. If both Al and Bob commit crimes, the police will only be able to devote half their resources to investigating each crime. Thus each criminal will have a probability .05 of being caught. If Al and Bob decide simultaneously whether

to commit a crime, we have a two-player game in normal form. Each player has two strategies: c (crime) and n (no crime). The payoff matrix is

		Bob	
		c	n
Al	c	$(.95b - .05f, .95b - .05f)$	$(.9b - .1f, 0)$
	n	$(0, .9b - .1f)$	$(0, 0)$

Use best response to show that if $9b < f < 19b$, there are two Nash equilibria: (c, c) (high crime) and (n, n) (low crime).

Thus if $f = 10b$, for example, current police resources may not deter crime. The game is of Stag Hunt type (Subsection 3.2.2). There is a low crime equilibrium, in which no one commits a crime, because the police could devote all their resources to catching the criminal. However, there is also a high crime equilibrium, in which crime pays, because police resources are not adequate to deal with a high level of crime.

(2) Again suppose two residents, Al and Bob, are considering committing crimes that if successful will produce a benefit b to the criminal. The police call in both Al and Bob and tell them: "We know you are both considering crimes. If one of you commits a crime, we will investigate it. However, if both of you commit a crime, we will only investigate Al's." The payoff matrix is now

		Bob	
		c	n
Al	c	$(.9b - .1f, b)$	$(.9b - .1f, 0)$
	n	$(0, .9b - .1f)$	$(0, 0)$

Explain the payoffs to the strategy profile (c, c). Then use best response to show that if $f > 9b$, no crimes are committed. Thus if $f = 10b$, for example, this policing strategy deters crime without any increase in police resources, whereas the strategy of investigating all crimes equally may not.

For examples of how this strategy has been put into effect, see [10].

(3) Now suppose m residents are considering committing crimes that if successful will produce a benefit b to the criminal. If a criminal is caught, he will suffer a punishment with value f. The policing level is r, $0 < r < 1$. This means that if k residents commit crimes, each will be caught with probability $\frac{r}{k}$. We view this situation as an m-player game in normal form. Each player has two strategies, c (crime) and n (no crime). The payoff to each player who chooses n is 0. If $k > 0$ players choose c, the payoff to

each is $\left(1-\frac{r}{k}\right)b-\frac{r}{k}f$. Assume that $b/(b+f) < r < mb/(b+f)$. Show: the only strategy profiles that are Nash equilibria are $(n,\ldots,n)$ and $(c,\ldots,c)$.

4.7.3 The Travails of Boss Gorilla 2.

This problem deals with a variant of the situation described in Section 4.5. Before he became boss of the Gorilla Band, Boss Gorilla had many encounters with other gorillas. Because of those encounters, it is well known that a certain fraction p of other gorillas can beat Boss Gorilla in a fight. The remaining fraction $1-p$ of gorillas will lose to Boss Gorilla in a fight. However, when a Visiting Gorilla appears, neither the Visiting Gorilla nor Boss Gorilla knows into which category the visitor falls.

We view this situation as a game with two players: Player 1, Boss Gorilla (B), and Player 2, Visiting Gorilla (V). Figure 4.10 shows the gorillas' possible moves and the payoffs, which come from Section 4.5. Now, however, there are two information sets, since neither gorilla knows whether Visiting Gorilla can beat Boss Gorilla; they only know the probability p.

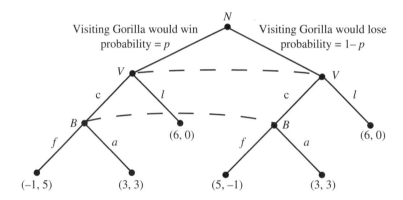

Figure 4.10. The Travails of Boss Gorilla with two information sets. Boss Gorilla (B) is Player 1, Visiting Gorilla (V) is Player 2. N = Nature.

Because of the information sets, Visiting Gorilla has just two strategies: challenge (c) or leave (l). Boss Gorilla also has just two strategies: fight if challenged (f) or acquiesce if challenged (a). In normal form, the game is represented by a 2×2 payoff matrix.

(1) Find the payoff matrix for the game. One way to do this is to find separately the payoff matrix when Visiting Gorilla would win and the payoff matrix when Visiting Gorilla would lose. Multiply the first matrix by p, multiply the second by $1-p$, and add.

(2) Use best response to find Nash equilibria in three cases: (a) $0 < p < \frac{1}{6}$, (b) $\frac{1}{6} < p < \frac{1}{3}$, and (c) $\frac{1}{3} < p < 1$.

4.7.4 Expert Opinion. We often face the following problem. Something of ours does not work properly (e.g., car, computer, body). We take it to an Expert (e.g., mechanic, computer repair person, doctor). The problem may be major or minor. The Expert studies the problem and diagnoses it as major or minor. We then must decide whether to follow the Expert's advice and do the repair:

- Expert's payoffs
  - Bill for major repair: M.
  - Bill for minor repair: m.
  - Boost to reputation from making correct diagnosis: B.
- Customer's payoffs
  - Value of getting major problem fixed: V.
  - Value of getting minor problem fixed: v.
  - Bill for major repair: $-M$.
  - Bill for minor repair: $-m$.

Half the time the problem is minor, and half the time it is major; everyone knows this.

If the problem is major, only a major repair will fix it. If the problem is minor, either a minor repair or a major repair will fix it.

We assume

$$V > M > v > m > 0 \quad \text{and} \quad B > m. \tag{4.1}$$

The game tree in Figure 4.11 illustrates the situation.

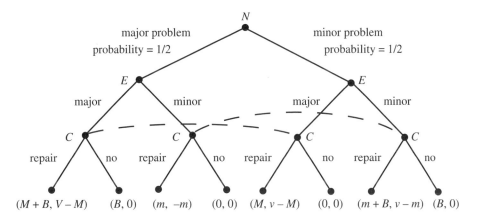

Figure 4.11. N = Nature, E = Expert, C = Customer. The Expert diagnoses a major or a minor problem; then the Customer decides whether to do the repair or not. The first payoff is to the Expert, the second is to the Customer.

The Customer has four strategies:

- rr: If Expert says problem is major, repair; if Expert says problem is minor, repair.
- rn: If Expert says problem is major, repair; if Expert says problem is minor, do not repair.
- nr: If Expert says problem is major, do not repair; if Expert says problem is minor, repair.
- nn: If Expert says problem is major, do not repair; if Expert says problem is minor, do not repair.

(1) Explain why the Expert should always diagnose a major problem as major. (What is the smallest payoff she gets by diagnosing a major problem as major, and what is the largest payoff she gets by diagnosing a major problem as minor?)

(2) Because of part (1), we assume the Expert always diagnoses a major problem as major. Therefore the Expert only has only two strategies that we need to consider:

- d (dishonest): diagnose every problem as major.

- h (honest): if the problem is major, diagnose it as major; if the problem is minor, diagnose it as minor.

To put this game in normal form, complete the following 2×4 payoff matrix, showing expected payoffs to both Expert and Customer:

		Customer			
		rr	rn	nr	nn
Expert	d				
	h				

(3) In addition to (4.1), assume $B > M$. Show that in this case the Expert's strategy h strictly dominates her strategy d. Find the unique Nash equilibrium.

(4) In addition to (4.1) assume $M > m + B$ and $V + v > 2M$. Use best response to find Nash equilibria.

4.7.5 The Value of College. This game has three players: very intelligent young people (V), less intelligent young people (L), and employers (E). The young people have two options: go to college (c) or not (n). When a young person applies for a job, an employer has two options: offer a high salary (h)

or offer a low salary (*l*). A very intelligent young person will only accept a high salary. A less intelligent young person will accept either salary.

- The Employer's payoffs are
    - Pay high salary: −8.
    - Pay low salary: −3.
    - Hire very intelligent young person: 12.
    - Hire less intelligent young person: 6.
- The very intelligent young person's payoffs are
    - Go to college: −1. (It teaches nothing important and is expensive, but it's easy.)
    - Get offered high salary: 8. (She takes the job and gets paid.)
    - Get offered low salary: 6. (The very intelligent young person will reject the offer and start her own business.)
- The less intelligent young person's payoffs are
    - Go to college: −5. (It teaches nothing important and is expensive, and it's really hard.)
    - Get offered high salary: 8.
    - Get offered low salary: 3. (The less intelligent young person will accept either offer and get paid.)

Half of young people are very intelligent, half are less intelligent; the employer knows this. When a young person applies for a job, the employer does not know how intelligent she is, only whether she has gone to college.

The game tree in Figure 4.12 illustrates the situation.

The young people each have two strategies: go to college (*c*) or not (*n*). An employer has four strategies:

- *hh*: Offer a high salary to all job applicants.
- *hl*: Offer a high salary to an applicant who has been to college, offer a low salary to an applicant who has not been to college.
- *lh*: Offer a low salary to an applicant who has been to college, offer a high salary to an applicant who has not been to college.
- *ll*: Offer a low salary to all job applicants.

Analogously to what was done in Section 4.5, we analyze this game by viewing it as a two-player game in which the players are the young person and the employer. Nature first decides the type of the young person, very intelligent or less intelligent, and then a game is played between the young person and the employer. The game tree for the two-player game is the one

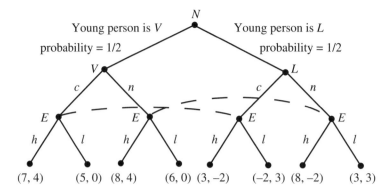

Figure 4.12. N = Nature, V = very intelligent young person, L = less intelligent young person, E = employer, c = college, n = no college, h = offer high salary, l = offer low salary. The first payoff is to the young person, and the second is to the employer.

shown in Figure 4.12, except that the nodes labeled V and L should both be labeled Y (young person). The young person is assumed to know her own type, so she can use different strategies in the two cases.

When the two-player game is converted to normal form, the employer still has the four strategies given above, but the young person now also has four strategies:

- cc: Always go to college.
- cn: Go to college if very intelligent, do not go to college if less intelligent.
- nc: Do not go to college if very intelligent, go to college if less intelligent.
- nn: Never go to college.

(1) Explain briefly why a less intelligent person should not go to college. More precisely, explain briefly why the young person's strategy cn weakly dominates her strategy cc, and the young person's strategy nn weakly dominates her strategy nc. (What is the best payoff a less intelligent young person can get by going to college, and what is the worst payoff she gets by not going to college?)

(2) Eliminate the young person's weakly dominated strategies cc and nc. Complete the following 2×4 payoff matrix for payoffs in the two-player game:

		hh	hl	lh	ll
			E		
Y	cn				
	nn				

(3) Use best response to find Nash equilibria. (There are two Nash equilibria. If you look at the corresponding Nash equilibria of the three-player game, you will see that they are equally good for the less intelligent young people, but one is better for both very intelligent young people and employers.)

Chapter 5

Mixed strategy Nash equilibria

Even simple games can fail to have Nash equilibria in the sense we have so far discussed. Sometimes it is best to mix your actions unpredictably. For example, a tennis player wants to serve sometimes to her opponent's forehand and sometimes to her backhand in a random manner that her opponent can't predict. But what fraction of her serves should go to each? This depends on the skills of both players.

5.1 Mixed strategy Nash equilibria

In tennis, the player serving can serve to her opponent's backhand or forehand. The player receiving the serve can anticipate a serve to her backhand or a serve to her forehand.

For a certain pair of tennis players, the probability that the serve is returned is given by the following table:

		Receiver anticipates serve to	
		backhand	forehand
Server serves to	backhand	.6	.2
	forehand	.3	.9

We regard this as a game in normal form. The payoff to the receiver is the fraction of serves she returns; the payoff to the server is the fraction of serves that are not returned. Thus the payoff matrix is

		Receiver anticipates serve to	
		backhand	forehand
Server serves to	backhand	(.4, .6)	(.8, .2)
	forehand	(.7, .3)	(.1, .9)

You can check using best response that there are no Nash equilibria in the sense we have discussed.

It is easy to understand why this game has no equilibria. If I plan to serve to your forehand, your best response is to anticipate a serve to your forehand. But if you anticipate a serve to your forehand, my best response is to serve to your backhand. But if I plan serve to your backhand, your best response is to anticipate a serve to your backhand. But if you anticipate a serve to your backhand, my best response is to serve to your forehand. We are back where we started from, without having found an equilibrium!

Does game theory have any suggestions for these players? Yes: the suggestion to the server is to mix her serves randomly, and the suggestion to the receiver is to mix her expectations randomly. What fraction of the time should the server serve to the forehand, and what fraction of the time should the receiver anticipate a serve to her forehand? To answer this question we must develop the idea of a mixed strategy.

Consider a game in normal form with players $1, \ldots, n$ and corresponding strategy sets $S_1, \ldots, S_n$, all finite. Suppose that for each i, Player i's strategy set consists of k_i strategies, which we denote $s_{i1}, \ldots, s_{ik_i}$. A *mixed strategy* σ_i for Player i consists of using strategy s_{i1} with probability p_{i1}, strategy s_{i2} with probability $p_{i2}, \ldots$, strategy s_{ik_i} with probability p_{ik_i}. Of course, each $p_{ij} \geqslant 0$, and $\sum_{j=1}^{k_i} p_{ij} = 1$. Formally, $\sigma_i = \sum_{j=1}^{k_i} p_{ij} s_{ij}$.

If $p_{ij} > 0$, we say that the pure strategy s_{ij} is *active* in the mixed strategy σ_i.

A mixed strategy σ_i is called *pure* if only one pure strategy is active (i.e., if one p_{ij} is 1 and all the rest are 0). We can denote by s_{ij} the pure strategy of Player i that uses s_{ij} with probability 1 and her other strategies with probability 0. Until this chapter we have only discussed pure strategies.

We try whenever possible to avoid double subscripting. Thus we often denote a strategy of Player i by s_i, and the associated probability by p_{s_i}. This requires summing over all $s_i \in S_i$ instead of summing from $j = 1$ to k_i. Thus a mixed strategy of Player i is written

$$\sigma_i = \sum_{\text{all } s_i \in S_i} p_{s_i} s_i. \tag{5.1}$$

If each player chooses a mixed strategy, we get a *mixed strategy profile* $(\sigma_1, \ldots, \sigma_n)$.

Recall that if each player chooses a pure strategy, we get a *pure strategy profile* $(s_1, \ldots, s_n)$. Recall that associated with each pure strategy profile $(s_1, \ldots, s_n)$ is a payoff to each player; the payoff to Player i is denoted $\pi_i(s_1, \ldots, s_n)$.

Suppose that

- Player 1's mixed strategy σ_1 uses her strategy s_1 with probability p_{s_1},
- Player 2's mixed strategy σ_2 uses her strategy s_2 with probability p_{s_2},
  ⋮
- Player n's mixed strategy σ_n uses her strategy s_n with probability p_{s_n}.

We assume that the players independently choose pure strategies to use. Then, if the players use the mixed strategy profile $(\sigma_1, \ldots, \sigma_n)$, the probability that the pure strategy profile $(s_1, \ldots, s_n)$ occurs is the product $p_{s_1} p_{s_2} \cdots p_{s_n}$. Thus the expected payoff to Player i is

$$\pi_i(\sigma_1, \ldots, \sigma_n) = \sum_{\text{all } (s_1, \ldots, s_n)} p_{s_1} p_{s_2} \cdots p_{s_n} \pi_i(s_1, \ldots, s_n). \tag{5.2}$$

Let σ denote the mixed strategy profile $(\sigma_1, \ldots, \sigma_n)$. Suppose in σ we replace the ith player's mixed strategy σ_i by another of her mixed strategies, say, τ_i. We denote the resulting mixed strategy profile by (τ_i, σ_{-i}). This notation is analogous to that introduced in Section 3.1.

If Player 1 uses a pure strategy s_1, (5.2) yields for Player 1's payoff

$$\pi_1(s_1, \sigma_2 \ldots, \sigma_n) = \sum_{\text{all } (s_2, \ldots, s_n)} p_{s_2} \cdots p_{s_n} \pi_1(s_1, \ldots, s_n). \tag{5.3}$$

Equation (5.2) for Player 1's payoff now can be rewritten as

$$\pi_1(\sigma_1, \ldots, \sigma_n) = \sum_{\text{all } s_1 \in S_1} p_{s_1} \sum_{\text{all } (s_2, \ldots, s_n)} p_{s_2} \cdots p_{s_n} \pi_1(s_1, \ldots, s_n). \tag{5.4}$$

Using (5.3), equation (5.4) can be written

$$\pi_1(\sigma_1, \ldots, \sigma_n) = \sum_{\text{all } s_1 \in S_1} p_{s_1} \pi_1(s_1, \sigma_2 \ldots, \sigma_n) = \sum_{\text{all } s_1 \in S_1} p_{s_1} \pi_1(s_1, \sigma_{-1}). \tag{5.5}$$

More generally, for Player i,

$$\pi_i(\sigma_1, \ldots, \sigma_n) = \sum_{\text{all } s_i \in S_i} p_{s_i} \pi_i(s_i, \sigma_{-i}). \tag{5.6}$$

In words, the payoff to Player i from using strategy σ_i against the other players' mixed strategies is just a weighted average of her payoffs from using her pure strategies against their mixed strategies, where the weights are the probabilities in her strategy σ_i.

A mixed strategy profile $(\sigma_1^*, \ldots, \sigma_n^*)$ is a *mixed strategy Nash equilibrium* if no single player can improve her own payoff by changing her strategy:

- For every mixed strategy σ_1 of Player 1,

$$\pi_1(\sigma_1^*, \sigma_2^*, \ldots, \sigma_n^*) \geqslant \pi_1(\sigma_1, \sigma_2^*, \ldots, \sigma_n^*).$$

- For every mixed strategy σ_2 of Player 2,

$$\pi_2(\sigma_1^*, \sigma_2^*, \sigma_3^*, \ldots, \sigma_n^*) \geqslant \pi_2(\sigma_1^*, \sigma_2, \sigma_3^*, \ldots, \sigma_n^*).$$

$\vdots$

- For every mixed strategy σ_n of Player n,

$$\pi_n(\sigma_1^*, \ldots, \sigma_{n-1}^*, \sigma_n^*) \geqslant \pi_n(\sigma_1^*, \ldots, \sigma_{n-1}^*, \sigma_n).$$

More compactly, a mixed strategy profile $\sigma^* = (\sigma_1^*, \ldots, \sigma_n^*)$ is a Nash equilibrium if, for each $i = 1, \ldots, n$, $\pi_i(\sigma^*) \geqslant \pi_i(\sigma_i, \sigma_{-i}^*)$ for every mixed strategy σ_i of Player i.

Theorem 5.1. *Nash's Existence Theorem. If, in an n-player game in normal form, each player's strategy set is finite, then the game has at least one mixed strategy Nash equilibrium.*

John Nash was awarded the Nobel Prize in Economics in 1994 largely for discovering this theorem. We shall not give the proof, which uses mathematical ideas beyond the scope of this course.

The definition of a mixed strategy Nash equilibrium implicitly assumes that the lottery with the higher expected utility is preferred. Therefore it should only be used in situations where the expected utility principle (Section 4.1) can reasonably be expected to hold.

The next result gives a characterization of Nash equilibria that is very useful in finding them.

Theorem 5.2. *Fundamental Theorem of Nash Equilibria. The mixed strategy profile $\sigma = (\sigma_1, \ldots, \sigma_n)$ is a mixed strategy Nash equilibrium if and only if the following two conditions are satisfied for every $i = 1, \ldots, n$.*

(1) If the strategies s_i and s_i' are both active in σ_i, then $\pi_i(s_i, \sigma_{-i}) = \pi_i(s_i', \sigma_{-i})$.
(2) If the strategy s_i is active in σ_i and the strategy s_i' is not active in σ_i, then $\pi_i(s_i, \sigma_{-i}) \geqslant \pi_i(s_i', \sigma_{-i})$.

This theorem just says that mixed strategy Nash equilibria are characterized by the following property: each player's active strategies are all best responses to the profile of the other players' mixed strategies, where "best response" means best response among pure strategies.

Proof. First, suppose $\sigma = (\sigma_1, \ldots, \sigma_n)$ is a mixed strategy Nash equilibrium with σ_i given by (5.1). We will show that conditions (1) and (2) hold.

Suppose that in Player i's strategy σ_i, her pure strategies s_i and s_i' are both active, with probabilities $p_{s_i} > 0$ and $p_{s_i'} > 0$, respectively. Could it be that $\pi_i(s_i, \sigma_{-i}) < \pi_i(s_i', \sigma_{-i})$? Look at (5.6). If this were the case, then Player i could switch to a new strategy τ_i that differs from σ_i only in that the pure strategy s_i is not used at all, but the pure strategy s_i' is used with probability $p_{s_i} + p_{s_i'}$. This would increase Player i's payoff, so $\sigma = (\sigma_1, \ldots, \sigma_n)$ would not be a mixed strategy Nash equilibrium. Similarly, it is not possible that $\pi_i(s_i, \sigma_{-i}) > \pi_i(s_i', \sigma_{-i})$. Hence condition (1) must hold.

Now suppose that in Player i's strategy σ_i, s_i is active, with probabiliy $p_{s_i} > 0$, and s_i' is not active. Could it be that $\pi_i(s_i, \sigma_{-i}) < \pi_i(s_i', \sigma_{-i})$? If this were the case, then Player i could switch to a new strategy τ_i that differs from σ_i only in that the pure strategy s_i is not used at all, but the pure strategy s_i' is used with probability p_{s_i}. From (5.6), this would increase Player i's payoff, so $\sigma = (\sigma_1, \ldots, \sigma_n)$ would not be a mixed strategy Nash equilibrium. Hence (2) must also hold.

To prove the converse, suppose $\sigma = (\sigma_1, \ldots, \sigma_n)$ is a mixed strategy profile, with σ_i given by (5.1), that satisfies conditions (1) and (2). We will show that σ is a mixed strategy Nash equilibrium.

Since conditions (1) and (2) hold, for Player i there is a number K such that $\pi_i(s_i, \sigma_{-i}) = K$ for each of Player i's active stragties s_i, and $\pi_i(s_i, \sigma_{-i}) \leqslant K$ for each of Player i's inactive stragties s_i. Let S_i^* denote the set of Player i's active strategies. We first note that $\pi_i(\sigma_1, \ldots, \sigma_n) = K$, because

$$\pi_i(\sigma_1, \ldots, \sigma_n) = \sum_{\text{all } s_i \in S_i} p_{s_i} \pi_i(s_i, \sigma_{-i}) = \sum_{s_i \in S_i^*} p_{s_i} \pi_i(s_i, \sigma_{-i}) = \sum_{s_i \in S_i^*} p_{s_i} K = K.$$

$$(5.7)$$

The first equality follows from (5.6). The second equality holds because only the p_{s_i} with $s_i \in S_i^*$ are nonzero. The last equality holds because $\sum_{s_i \in S_i^*} p_{s_i} = 1$.

Let τ_i be any other strategy for Player i,

$$\tau_i = \sum_{\text{all } s_i \in S_i} q_{s_i} s_i, \quad \text{all } q_{s_i} \geqslant 0, \quad \sum q_{s_i} = 1.$$

Then from (5.6),

$$\pi_i(\tau_i, \sigma_{-i}) = \sum_{\text{all } s_i \in S_i} q_{s_i} \pi_i(s_i, \sigma_{-i}).$$

Since each $\pi_i(s_i, \sigma_{-i}) \leqslant K$, each $q_{s_i} \geqslant 0$, and $\sum q_{s_i} = 1$, this sum is at most K. Therefore it is at most $\pi_i(\sigma_1, \ldots, \sigma_n) = K$. Thus Player i cannot improve her payoff by switching to τ_i. It follows that $(\sigma_1, \ldots, \sigma_n)$ is a Nash equilibrium. $\square$

In the remainder of this section we give some easy consequences of the Fundamental Theorem of Nash Equilibria and its proof, and we comment on the relationship between mixed strategy Nash equilibria and iterated elimination of dominated strategies.

The calculation (5.7) is so important that we record it, and a generalization, as the following theorem.

Theorem 5.3. *Let* $\sigma = (\sigma_1, \ldots, \sigma_n)$ *be a mixed strategy Nash equilibrium with* σ_i *given by (5.1). Let* $S_i^* = \{s_i \colon p_{s_i} > 0\}$ *be the set of Player i's active strategies. For* $s_i \in S_i^*$, *let* $\pi_i(s_i, \sigma_{-i}) = K$. *(According to the Fundamental Theorem, all the payoffs* $\pi_i(s_i, \sigma_{-i})$ *with* $s_i \in S_i^*$ *are equal.) Then*

(1) $\pi_i(\sigma) = K$.
(2) *More generally, let* τ_i *be any combination of Player i's active strategies, that is,* $\tau_i = \sum_{s_i \in S_i^*} q_{s_i} s_i$ *with all* $q_{s_i} \geqslant 0$ *and* $\sum q_{s_i} = 1$. *Then* $\pi_i(\tau_i, \sigma_{-i}) = K$.

Proof. We already noticed conclusion (1) during the proof of the previous theorem. To prove the more general result (2), we calculate

$$\pi_i(\tau_i, \sigma_{-i}) = \sum_{\text{all } s_i \in S_i} q_{s_i} \pi_i(s_i, \sigma_{-i}) = \sum_{s_i \in S_i^*} q_{s_i} \pi_i(s_i, \sigma_{-i}) = \sum_{s_i \in S_i^*} q_{s_i} K = K.$$

The first equality follows from (5.6). The second equality holds because only the q_{s_i} with $s_i \in S_i^*$ are nonzero. The last equality holds because $\sum q_{s_i} = 1$. □

At a mixed strategy Nash equilibrium $(\sigma_1, \ldots, \sigma_n)$, σ_i is a best response to the profile of the other players' mixed strategies. Theorem 5.3 says that all of Player i's active strategies are also best responses to the profile of the other players' mixed strategies, and in fact any combination of Player i's active strategies is a best response to the profile of the other players' mixed strategies.

Theorem 5.4. *If* $s^* = (s_1^*, \ldots, s_n^*)$ *is a profile of pure strategies, then* s^* *is a mixed strategy Nash equilibrium if and only if* s^* *is a Nash equilibrium in the sense of Section 3.1.*

Proof. Suppose $s^* = (s_1^*, \ldots, s_n^*)$ is a profile of pure strategies that is a mixed strategy Nash equilibrium. Consider Player i and one of her strategies s_i other than s_i^*. Since s_i is not active, the second part of the Fundamental Theorem says that $\pi_i(s_i^*, s_{-i}^*) \geqslant \pi_i(s_i, s_{-i}^*)$, so s^* is a Nash equilibrium in the sense of Section 3.1.

On the other hand, suppose $s^* = (s_1^*, \ldots, s_n^*)$ is a profile of pure strategies that is a Nash equilibrium in the sense of Section 3.1. Then condition (1) of the Fundamental Theorem is automatically satisfied, since each player has only one active strategy. Condition (2) is an immediate consequence of the definition of Nash equilibrium in Section 3.1. Since both conditions hold, s^* is a mixed strategy Nash equilibrium. □

A mixed strategy profile $\sigma^* = (\sigma_1^*, \ldots, \sigma_n^*)$ is a *strict mixed strategy Nash equilibrium* if, for each $i = 1, \ldots, n$, $\pi_i(\sigma^*) > \pi_i(\sigma_i, \sigma_{-i}^*)$ for every mixed strategy $\sigma_i \neq \sigma_i^*$ of Player i.

The following result is a consequence of the previous two theorems.

Theorem 5.5. $\sigma^* = (\sigma_1^*, \ldots, \sigma_n^*)$ *is a strict mixed strategy Nash equilibrium if and only if (i) each σ_i^* is a pure strategy, and (ii) σ^* is a strict Nash equilibrium in the sense of Section 3.1.*

For most two-player games, at each Nash equilibrium, both players use the same number of active pure strategies. Thus in two-player games, one can begin by looking for Nash equilibria in which each player uses one active pure strategy, then Nash equilibria in which each player uses two active pure strategies, and so forth. In most two-player games, this procedure not only finds all mixed strategy Nash equilibria; it also yields as a by-product a proof that there are no Nash equilibria in which the two players use different numbers of active pure strategies.

Finally, we comment on the relationship between mixed strategy Nash equilibria and iterated elimination of dominated strategies. Suppose that, as in Section 3.6, we do iterated elimination of weakly dominated pure strategies on a game G in normal form. Let H be the reduced game that results. Then each mixed strategy Nash equilibrium of H is also a mixed strategy Nash equilibrium of G. If we do iterated elimination of strictly dominated pure strategies, then each strategy that is eliminated is not part of any mixed strategy Nash equilibrium of G.

5.2 Tennis

Recall the game of tennis described in Section 5.1, with the payoff matrix

		Receiver anticipates serve to	
		backhand	forehand
Server serves to	backhand	(.4, .6)	(.8, .2)
	forehand	(.7, .3)	(.1, .9)

We shall use the Fundamental Theorem of Nash Equilibria, Theorem 5.2, to find all mixed strategy Nash equilibria in this game.

Suppose the receiver uses her two strategies with probabilities p and $1 - p$, and the server uses her strategies with probabilities q and $1 - q$. It is helpful to write these probabilities next to the payoff matrix as follows:

			Receiver anticipates serve to	
			q	$1 - q$
			backhand	forehand
Server serves to	p	backhand	(.4, .6)	(.8, .2)
	$1 - p$	forehand	(.7, .3)	(.1, .9)

We shall look for a mixed strategy Nash equilibria $(pb + (1 - p)f, qb + (1 - q)f)$.

In accordance with the advice in the previous section, we shall first look for equilibria in which both players use one active pure strategy, then look for equilibria in which both players use two active pure strategies.

1. Suppose both players use one active pure strategy. Then we would have a pure strategy Nash equilibrium. We find these using best response. You checked in Section 5.1 that there aren't any.

2. Suppose both players use two active pure strategies. Then $0 < p < 1$ and $0 < q < 1$. Since both of Player 2's pure strategies b and f are active, according to the Fundamental Theorem, each gives the same payoff to Player 2 against Player 1's mixed strategy $pb + (1 - p)f$:

$$\pi_2(pb + (1 - p)f, b) = \pi_2(pb + (1 - p)f, f),$$

or

$$.6p + .3(1 - p) = .2p + .9(1 - p).$$

Solving this equation for p, we find that $p = .6$.

Similarly, since both of Player 1's pure strategies b and f are active, each gives the same payoff to Player 1 against Player 2's mixed strategy $qb + (1 - q)f$:

$$\pi_1(b, qb + (1 - q)f) = \pi_1(f, qb + (1 - q)f),$$

or

$$.4q + .8(1 - q) = .7q + .1(1 - q).$$

Solving this equation for q, we find that $q = .7$.

We conclude that $(.6b + .4f, .7b + .3f)$ satisfies the equality criterion for a mixed strategy Nash equilibrium. Since there are no unused pure strategies, there is no inequality criterion to check. Therefore we have found a mixed strategy Nash equilibrium in which both players have two active pure strategies.

Note that in the course of finding this Nash equilibrium, we actually did more:

(1) We showed that if both of Player 2's pure strategies are active at a Nash equilibrium, then Player 1's strategy must be $.6b + .4f$. Hence there are no Nash equilibria in which Player 2 uses two pure strategies but Player 1 uses only one pure strategy.

(2) Similarly, we showed that if both of Player 1's pure strategies are active at a Nash equilibrium, then Player 2's strategy must be $.7b + .3f$. Hence there are no Nash equilibria in which Player 1 uses two pure strategies but Player 2 uses only one pure strategy.

This is an example of how, in the course of finding mixed strategy Nash equilibria for a two-player game in which both players use the same number of pure strategies, one usually shows as a by-product that there are no Nash equilibria in which the two players use different numbers of pure strategies.

5.3 Other ways to find mixed strategy Nash equilibria

Here are two ways to find mixed strategy Nash equilibria in the previous problem without using the Fundamental Theorem. Both may be useful in other problems.

5.3.1 Differentiating the payoff functions.
In the tennis problem, there are two payoff functions, π_1 and π_2. Since Player 1's strategy is determined by the choice of p and Player 2's by the choice of q, we may regard π_1 and π_2 as functions of (p, q) with $0 \leqslant p \leqslant 1$ and $0 \leqslant q \leqslant 1$. From the payoff matrix, we have

$$\pi_1(p, q) = .4pq + .8p(1 - q) + .7(1 - p)q + .1(1 - p)(1 - q),$$
$$\pi_2(p, q) = .6pq + .2p(1 - q) + .3(1 - p)q + .9(1 - p)(1 - q).$$

Suppose (p, q) is a mixed strategy Nash equilibrium with $0 < p < 1$ and $0 < q < 1$. Then the definition of mixed strategy Nash equilibrium implies that

$$\frac{\partial \pi_1}{\partial p}(p, q) = 0 \quad \text{and} \quad \frac{\partial \pi_2}{\partial q}(p, q) = 0.$$

Therefore

$$\frac{\partial \pi_1}{\partial p}(p, q) = .4q + .8(1 - q) - .7q - .1(1 - q) = .7 - q = 0,$$

$$\frac{\partial \pi_2}{\partial q}(p, q) = .6p - .2p + .3(1 - p) - .9(1 - p) = -.6 + p = 0.$$

We see that $(p, q) = (.6, .7)$.

5.3.2 Best-response correspondences. From the calculation of partial derivatives above, we have

$$\frac{\partial \pi_1}{\partial p}(p,q) = \begin{cases} + & \text{if } q < .7, \\ 0 & \text{if } q = .7, \\ - & \text{if } q > .7, \end{cases} \qquad \frac{\partial \pi_2}{\partial q}(p,q) = \begin{cases} - & \text{if } p < .6, \\ 0 & \text{if } p = .6, \\ + & \text{if } p > .6. \end{cases}$$

These partial derivatives tell us each player's best response to all strategies of her opponent. For Player 1:

- If Player 2 chooses q with $0 \leqslant q < .7$, Player 1 observes that her own payoff is an increasing function of p. Hence her best response is $p = 1$.
- If Player 2 chooses $q = .7$, Player 1 observes that her own payoff will be the same whatever p she chooses. Hence she can choose any p between 0 and 1.
- If Player 2 chooses q with $.7 < q \leqslant 1$, Player 1 observes that her own payoff is a decreasing function of p. Hence her best response is $p = 0$.

Player 1's best-response correspondence $B_1(q)$ is graphed in Figure 5.1, along with Player 2's best-response correspondence $B_2(p)$. Note that $B_1(.7)$ is the set $0 \leqslant p \leqslant 1$ and $B_2(.6)$ is the set $0 \leqslant q \leqslant 1$. Points of intersection of the two graphs are Nash equilibria. In this case, the only point of intersection of the two graphs is $(p,q) = (.6, .7)$.

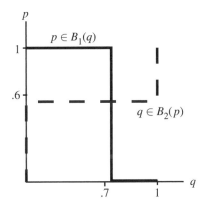

Figure 5.1. Graphs of best-response correspondences in the game of tennis. The only point in the intersection of the two graphs is $(p,q) = (.6, .7)$.

5.4 One-card Two-round Poker

We now play a simplified version of poker with a deck of two cards, one high (H) and one low (L). There are two players. Play proceeds as follows:

(1) Each player puts $2 into the pot.
(2) Player 1 is dealt one card, chosen by chance. He looks at it. He either bets $2 or he folds. If he folds, Player 2 gets the pot. If he bets:
(3) Player 2 either bets $2 or he folds. If he folds, Player 1 gets the pot. If he bets:
(4) Player 1 either bets $2 or he folds. If he folds, Player 2 gets the pot. If he bets:
(5) Player 2 either bets $2 or he folds. If he folds, Player 1 gets the pot. If he bets, Player 1 shows his card. If it is H, Player 1 wins the pot. If it is L, Player 2 wins the pot.

The game tree is shown in Figure 5.2.

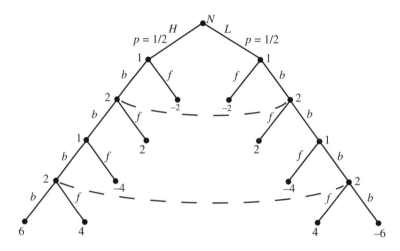

Figure 5.2. One-card two-round poker. Nodes in the same information set are linked by a dashed line. Only payoffs to Player 1 are shown; payoffs to Player 2 are opposite (the negative of Player 1's payoffs).

Player 2 has three pure strategies:

- bb: the first time Player 1 bets, respond by betting; the second time Player 1 bets, respond by betting.
- bf: the first time Player 1 bets, respond by betting; the second time Player 1 bets, respond by folding.
- f: the first time Player 1 bets, respond by folding.

(Actually, f is not a strategy in the sense that we have defined the term, since it does not specify what Player 2 will do after Player 1 bets for a second time. We have not specified this choice because, if Player 2 correctly does what he intends and folds after one bet by Player 1, the question of what to do after a second bet by Player 1 will not arise. Thus this choice does not affect the payoffs that we calculate.)

To describe Player 1's pure strategies, we first note that if Player 1 is dealt the high card, he has three options:

- bb: bet; if Player 2 responds by betting, bet again.
- bf: bet; if Player 2 responds by betting, fold.
- f: fold.

If Player 1 is dealt the low card, he has the same three options. Thus Player 1 has nine pure strategies: choose one option to use if dealt the high card, and one to option to use if dealt the low card.

The payoff matrix for this game is 9×3. If we draw it, we will quickly see that six of Player 1's pure strategies are weakly dominated: every strategy of Player 1 that does not use the option bb when dealt the high card is weakly dominated by the corresponding strategy that does use this option. This is obviously correct: if Player 1 is dealt the high card, he will certainly gain a positive payoff if he continues to bet, and will certainly suffer a negative payoff if he ever folds.

We therefore eliminate six of Player 1's strategies and obtain a reduced 3×3 game. In the reduced game, we denote Player 1's strategies by bb, bf, and f. The notation represents the option Player 1 uses if dealt the low card; if dealt the high card, he uses the option bb. By Theorem 3.1, any Nash equilibria of the reduced game are also Nash equilibria of the full game.

If the card dealt is high, the payoffs are

		Player 2		
		bb	bf	f
	bb	$(6, -6)$	$(4, -4)$	$(2, -2)$
Player 1	bf	$(6, -6)$	$(4, -4)$	$(2, -2)$
	f	$(6, -6)$	$(4, -4)$	$(2, -2)$

If the card dealt is low, the payoffs are

		Player 2		
		bb	bf	f
	bb	$(-6, 6)$	$(4, -4)$	$(2, -2)$
Player 1	bf	$(-4, 4)$	$(-4, 4)$	$(2, -2)$
	f	$(-2, 2)$	$(-2, 2)$	$(-2, 2)$

The payoff matrix for the game is $\frac{1}{2}$ times the first matrix plus $\frac{1}{2}$ times the second:

			Player 2		
		bb	***bf***	***f***
	bb	$(0,0)$	$(4,-4)$	$(2,-2)$
Player 1	***bf***	$(1,-1)$	$(0,0)$	$(2,-2)$
	f	$(2,-2)$	$(1,-1)$	$(0,0)$

We shall look for a mixed strategy Nash equilibrium (σ_1, σ_2), with $\sigma_1 = p_1 bb + p_2 bf + p_3 f$ and $\sigma_2 = q_1 bb + q_2 bf + q_3 f$:

				Player 2		
			q_1	q_2	q_3
			bb	***bf***	***f***
	p_1	***bb***	$(0,0)$	$(4,-4)$	$(2,-2)$
Player 1	p_2	***bf***	$(1,-1)$	$(0,0)$	$(2,-2)$
	p_3	***f***	$(2,-2)$	$(1,-1)$	$(0,0)$

We should consider three possibilities:

(1) Both players use a pure strategy.
(2) Both players use exactly two active pure strategies.
(3) Both players use exactly three active pure strategies.

One easily deals with the first case using best response: there are no pure strategy Nash equilibria. The second case divides into nine subcases, since each player has three ways to choose his two active strategies. For now we ignore these possibilities; we will return to them in the Section 5.5.

In the third case, we assume that all p_i and q_i are positive. Since all q_i are positive, each of Player 2's pure strategies gives the same payoff to Player 2 against Player 1's mixed strategy σ_1. These three payoffs are:

$$\pi_2(\sigma_1, bb) = p_1(0) + p_2(-1) + p_3(-2),$$
$$\pi_2(\sigma_1, bf) = p_1(-4) + p_2(0) + p_3(-1),$$
$$\pi_2(\sigma_1, f) = p_1(-2) + p_2(-2) + p_3(0).$$

The fact that these three quantities must be equal yields two independent equations. For example, one can use $\pi_2(\sigma_1, bb) = \pi_2(\sigma_1, f)$ and $\pi_2(\sigma_1, bf) = \pi_2(\sigma_1, f)$:

$$p_1(0) + p_2(-1) + p_3(-2) = p_1(-2) + p_2(-2) + p_3(0),$$
$$p_1(-4) + p_2(0) + p_3(-1) = p_1(-2) + p_2(-2) + p_3(0).$$

Simplifying, we have

$$2p_1 + p_2 - 2p_3 = 0,$$
$$-2p_1 + 2p_2 - p_3 = 0.$$

A third equation is given by

$$p_1 + p_2 + p_3 = 1.$$

These three equations in the three unknowns (p_1, p_2, p_3) can be solved to yield the solution

$$(p_1, p_2, p_3) = (\tfrac{1}{5}, \tfrac{2}{5}, \tfrac{2}{5}).$$

Had any p_i failed to lie strictly between 0 and 1, we would discard the possibility that there is a Nash equilibrium in which both players use three active pure strategies.

(One way to use the third equation is to use it to substitute $p_3 = 1 - p_1 - p_2$ in the first two equations. This is analogous to how we solved the tennis problem.)

Similarly, since all p_i are positive, each of Player 1's three pure strategies gives the same payoff to Player 1 against Player 2's mixed strategy σ_2. This observation leads to three equations in the three unknowns (q_1, q_2, q_3), which can be solved to yield

$$(q_1, q_2, q_3) = (\tfrac{8}{15}, \tfrac{2}{15}, \tfrac{1}{3}).$$

When you find a Nash equilbrium of a game in extensive form in which players move several times, it is useful to translate it into plans for the play of the game. In this case, the mixed strategy Nash equilibrium we have found yields the following plans:

- Player 1: if dealt the high card, bet at every opportunity. If dealt the low card:
    - Bet with probability $\tfrac{3}{5}$, fold with probability $\tfrac{2}{5}$.
    - If you get to bet a second time, bet with probability $\tfrac{1}{3}$, fold with probability $\tfrac{2}{3}$.
- Player 2:
    - If Player 1 bets, bet with probability $\tfrac{2}{3}$, fold with probability $\tfrac{1}{3}$.
    - If you get to bet a second time, bet with probability $\tfrac{4}{5}$, fold with probability $\tfrac{1}{5}$.

Player 1's strategy includes a lot of bluffing (betting when he has the low card)! Because Player 1 bluffs so much, it is rational for Player 2 to bet a lot, even though he has no idea what the situation is.

Note that in searching for Nash equilibria in which both players use all three of their pure strategies, we in fact showed:

(1) If all three of Player 2's pure strategies are active at a Nash equilibrium, then Player 1's strategy must be $(p_1, p_2, p_3) = (\tfrac{1}{5}, \tfrac{2}{5}, \tfrac{2}{5})$. Hence there are

no Nash equilibria in which Player 1 uses only one or two pure strategies and Player 2 uses three pure strategies.

(2) If all three of Player 1's pure strategies are active at a Nash equilibrium, then Player 2's strategy must be $(q_1, q_2, q_3) = \left(\frac{8}{15}, \frac{2}{15}, \frac{1}{3}\right)$. Hence there are no Nash equilibria in which Player 1 uses three pure strategies but Player 2 uses only one or two pure strategies.

If the two players use these strategies, the expected payoff to each player is given by (5.2). For example, the expected payoff to Player 1 is

$$\pi_1(\sigma_1, \sigma_2) = p_1 q_1 \cdot 0 + p_1 q_2 \cdot 4 + p_1 q_3 \cdot 2 + p_2 q_1 \cdot 1$$
$$+ p_2 q_2 \cdot 0 + p_2 q_3 \cdot 2 + p_3 q_1 \cdot 2 + p_3 q_2 \cdot 1 + p_3 q_3 \cdot 0.$$

Substituting the values of the p_i and q_j that we have calculated, we find that the expected payoff to Player 1 is $\frac{6}{5}$. Alternatively, since (σ_1, σ_2) is a Nash equilibrium, by Theorem 5.3, the expected payoff to Player 1 is also given by any of the equal quantities $\pi_1(bb, \sigma_2)$, $\pi_1(bf, \sigma_2)$, or $\pi_1(f, \sigma_2)$, so we could calculate one of these instead. For example, $\pi_1(bb, \sigma_2) = q_1 \cdot 0 + q_2 \cdot 4 + q_3 \cdot 2 = \frac{2}{15} \cdot 4 + \frac{1}{3} \cdot 2 = \frac{6}{5}$.

Since the payoffs to the two players must add up to 0, the expected payoff to Player 2 is $-\frac{6}{5}$. In particular, we have found an arguably better strategy for Player 2 than always folding, which you might have suspected would be his best strategy. If Player 2 always folds and Player 1 uses his best response, which is to always bet, then Player 2's payoff is -2.

Tom Ferguson is a mathematician at UCLA, where he teaches game theory. His online game theory text is available at http://www.math.ucla.edu/~tom/ Game_Theory/Contents.html, and his home page is http://www.math.ucla .edu/~tom. On his home page you will find some articles applying game theory to poker that he wrote with his son, Chris Ferguson. Chris is a champion poker player, having won over $7 million. His Wikipedia page is http://en .wikipedia.org/wiki/Chris_Ferguson. A 2009 *New Yorker* article about Chris Ferguson, poker, and game theory is available at http://www.newyorker.com/ reporting/2009/03/30/090330fa_fact_wilkinson.

5.5 Two-player zero-sum games

One-card Two-round Poker is an example of a *two-player zero-sum game*. "Zero-sum" means that the two players' payoffs always add up to 0; if one player does better, the other must do worse. The Nash equilibria of two-player zero-sum games have several useful properties.

To explore these properties, we first define, for any two-player game in normal form, *maximin strategies*, *maximin payoffs*, and *maximin equilibria*.

If Player 1 uses strategy σ_1, her minimum possible payoff is

$$m_1(\sigma_1) = \min_{\sigma_2} \pi_1(\sigma_1, \sigma_2).$$

A *maximin strategy* for Player 1, denoted $\sigma_1^\dagger$, is a strategy that makes this minimum possible payoff as high as possible. Player 1's minimum possible payoff when she uses the strategy $\sigma_1^\dagger$ is her *maximin payoff $m_1^\dagger$*, given by

$$m_1^\dagger = m_1(\sigma_1^\dagger) = \max_{\sigma_1} m_1(\sigma_1) = \max_{\sigma_1} \min_{\sigma_2} \pi_1(\sigma_1, \sigma_2).$$

By using the strategy $\sigma_1^\dagger$, Player 1 guarantees herself a payoff of at least $m_1^\dagger$, no matter what Player 2 does. This is the highest payoff that Player 1 can guarantee herself.

Similarly, if Player 2 uses strategy σ_2, her minimum possible payoff is

$$m_2(\sigma_2) = \min_{\sigma_1} \pi_2(\sigma_1, \sigma_2).$$

A *maximin strategy* for Player 2, denoted $\sigma_2^\dagger$, is a strategy that makes this minimum possible payoff as high as possible. Player 2's minimum possible payoff when she uses the strategy $\sigma_2^\dagger$ is her *maximin payoff $m_2^\dagger$*, given by

$$m_2^\dagger = m_2(\sigma_2^\dagger) = \max_{\sigma_2} m_2(\sigma_2) = \max_{\sigma_2} \min_{\sigma_1} \pi_2(\sigma_1, \sigma_2).$$

By using the strategy $\sigma_2^\dagger$, Player 2 guarantees herself a payoff of at least $m_2^\dagger$, no matter what Player 1 does. This is the highest payoff that Player 2 can guarantee herself.

A strategy profile $(\sigma_1^\dagger, \sigma_2^\dagger)$, in which both players use maximin strategies, is called a *maximin equilibrium*. Each player's payoff is at least her maximin payoff $m_i^\dagger$.

For example, consider the Battle of Sexes in Subsection 3.2.4. Let

$$\sigma_1 = p \cdot \text{concert} + (1 - p) \cdot \text{wrestling}, \quad \sigma_2 = q \cdot \text{concert} + (1 - q) \cdot \text{wrestling}.$$

We shall represent a strategy profile by (p, q) instead of (σ_1, σ_2). Then

$$\pi_1(p, q) = 2 \cdot pq + 1 \cdot (1 - p)(1 - q), \quad \frac{\partial \pi_1}{\partial q} = 2 \cdot p - 1 \cdot (1 - p) = -1 + 3p.$$

Therefore

$$\frac{\partial \pi_1}{\partial q} = \begin{cases} - & \text{if } 0 \leqslant p < \frac{1}{3}, \\ 0 & \text{if } p = \frac{1}{3}, \\ + & \text{if } \frac{1}{3} < p \leqslant 1. \end{cases}$$

Hence

$$m_1(p) = \min_{0 \leqslant q \leqslant 1} \pi_1(p,q) = \begin{cases} \pi_1(p,1) & \text{if } 0 \leqslant p < \frac{1}{3}, \\ \text{any } \pi_1(p,q) & \text{if } p = \frac{1}{3}, \\ \pi_1(p,0) & \text{if } \frac{1}{3} < p \leqslant 1. \end{cases}$$

Therefore Player 1's minimium possible payoff when she uses the strategy p is

$$m_1(p) = \min_{0 \leqslant q \leqslant 1} \pi_1(p,q) = \begin{cases} 2p & \text{if } 0 \leqslant p < \frac{1}{3}, \\ \frac{2}{3} & \text{if } p = \frac{1}{3}, \\ 1-p & \text{if } \frac{1}{3} < p \leqslant 1. \end{cases}$$

If we graph $m_1(p)$, we see that it is maximum at $p = \frac{1}{3}$, and $\pi_1(\frac{1}{3},q) = \frac{2}{3}$ for any q. Therefore Player 1's maximin strategy is $p^\dagger = \frac{1}{3}$, and her maximin payoff is $m_1^\dagger = \frac{2}{3}$.

A similar calculation shows that Player 2's maximin strategy is $q^\dagger = \frac{2}{3}$, and her maximin payoff is $m_2^\dagger = \frac{2}{3}$. Therefore $(p^\dagger, q^\dagger) = (\frac{1}{3}, \frac{2}{3})$ is a maximin equilibrium. It is not, however, a Nash equilibrium; the Nash equilibria are $(0,0)$, $(1,1)$, and $(\frac{2}{3}, \frac{1}{3})$.

The situation for a two-player zero-sum game is much nicer.

Theorem 5.6. *For a two-player zero-sum game:*

(1) A strategy profile is a Nash equilibrium if and only if it is a maximin equilibrium.

(2) At any Nash equilibrium, both players get their maximin payoffs.

(3) Let $m_i^\dagger$ denote Player i's maximin payoff. Then $m_1^\dagger + m_2^\dagger = 0$.

Proof. To prove conclusion (1) of the theorem, first assume (σ_1^*, σ_2^*) is a Nash equilibrium. We will show that (σ_1^*, σ_2^*) is also a maximin equilibrium. Let σ_1' be any strategy of Player 1, and let σ_2' be a corresponding strategy of Player 2 that gives Player 1 her lowest possible payoff when she uses σ_1'. In other words, suppose

$$m_1(\sigma_1') = \pi_1(\sigma_1', \sigma_2') \leqslant \pi_1(\sigma_1', \sigma_2) \quad \text{for all } \sigma_2. \tag{5.8}$$

Then

$$m_1(\sigma_1') = \pi_1(\sigma_1', \sigma_2') \leqslant \pi_1(\sigma_1', \sigma_2^*) \leqslant \pi_1(\sigma_1^*, \sigma_2^*). \tag{5.9}$$

The first inequality holds because of (5.8). The second holds because (σ_1^*, σ_2^*) is a Nash equilibrium.

Now $m_1^\dagger$ is the maximum of the numbers $m_1(\sigma_1')$, so (5.9) implies that $m_1^\dagger \leqslant \pi_1(\sigma_1^*, \sigma_2^*)$. We will show that $\pi_1(\sigma_1^*, \sigma_2^*)$ is itself Player 1's lowest possible payoff when she uses the strategy σ_1^*. Thus σ_1^* is a maximin strategy for Player 1, and $\pi_1(\sigma_1^*, \sigma_2^*) = m_1^\dagger$. Similarly, σ_2^* is a maximin strategy for Player 2, and $\pi_2(\sigma_1^*, \sigma_2^*) = m_2^\dagger$. Thus (σ_1^*, σ_2^*) is a maximin equilibrium; in addition, we have conclusion (2) of the theorem. Since $\pi_1(\sigma_1^*, \sigma_2^*) + \pi_2(\sigma_1^*, \sigma_2^*) = 0$, it follows that $m_1^\dagger + m_2^\dagger = 0$, which is conclusion (3) of the theorem.

Note that we showed conclusion (3) of the theorem under the assumption that a Nash equilibrium exists. However, this is always true by Nash's Existence Theorem (Theorem 5.1).

To prove that $\pi_1(\sigma_1^*, \sigma_2^*)$ is Player 1's lowest possible payoff when she uses the strategy σ_1^*, we note that

$$\pi_2(\sigma_1^*, \sigma_2^*) \geqslant \pi_2(\sigma_1^*, \sigma_2) \quad \text{for all } \sigma_2, \tag{5.10}$$

because (σ_1^*, σ_2^*) is a Nash equilibrium. But then, for any σ_2,

$$\pi_1(\sigma_1^*, \sigma_2^*) = -\pi_2(\sigma_1^*, \sigma_2^*) \leqslant -\pi_2(\sigma_1^*, \sigma_2) = \pi_1(\sigma_1^*, \sigma_2).$$

The two equalities hold because the game is zero-sum, and the inequality follows from (5.10).

To complete the proof of conclusion (1) of Theorem 5.6, let $(\sigma_1^\dagger, \sigma_2^\dagger)$ be a maximin equilibrium. To show that $(\sigma_1^\dagger, \sigma_2^\dagger)$ is also a Nash equilibrium, let σ_1 and σ_2 be arbitrary strategies for Players 1 and 2, respectively. We must show that $\pi_1(\sigma_1, \sigma_2^\dagger) \leqslant \pi_1(\sigma_1^\dagger, \sigma_2^\dagger)$ and $\pi_2(\sigma_1^\dagger, \sigma_2) \leqslant \pi_2(\sigma_1^\dagger, \sigma_2^\dagger)$. We shall show the first inequality; the second is similar. Since $\sigma_2^\dagger$ is Player 2's maximin strategy,

$$m_2^\dagger = m_2(\sigma_2^\dagger) \leqslant \pi_2(\sigma_1, \sigma_2^\dagger). \tag{5.11}$$

Therefore

$$\pi_1(\sigma_1, \sigma_2^\dagger) = -\pi_2(\sigma_1, \sigma_2^\dagger) \leqslant -m_2^\dagger = m_1^\dagger = m_1(\sigma_1^\dagger) \leqslant \pi_1(\sigma_1^\dagger, \sigma_2^\dagger).$$

The first equality holds because we have a zero-sum game; the second because of conclusion (3) of the theorem, which has already been proved; and the third by the definition of $m_1^\dagger$. The first inequality follows from (5.11). The second holds by the definition of $m_1(\sigma_1^\dagger)$. □

Theorem 5.6 implies that all Nash equilibria of a two-player zero-sum game are obtained by pairing any maximin strategy of Player 1 with any maximin strategy of Player 2. In particular, Player 1's strategy from one Nash equilibrium may be paired with Player 2's strategy from a second Nash equilibrium to yield a third Nash equilibrium. All three Nash equilibria, along with any others, will give the same payoff to Player 1, and the same opposite payoff to Player 2.

In many problems, Theorem 5.6 can be used to rule out the existence of other Nash equilibria once you have found one. In One-card Two-round Poker, for example, we found a Nash equilibrium in which both players had three active pure strategies. We saw that our calculations to find this Nash equilibrium ruled out the existence of Nash equilibria in which Player 1 used one or two pure strategies and Player 2 used three pure strategies.

Now suppose there were a Nash equilibria in which Player 1 used one or two pure strategies and Player 2 used one or two pure strategies. Then we could pair Player 1's strategy from this Nash equilibrium with Player 3's strategy from the Nash equilibrium we found, thus obtaining a Nash equilibrium in which Player 1 used one or two pure strategies and Player 2 used three pure strategies. However, we just saw that there is no such Nash equilibrium.

We conclude that the Nash equilibrium we found for One-card Two-round Poker is the only one there is.

You may recall that when we were working on One-card Two-round Poker, we did not investigate the nine types of possible Nash equilibria in which each player used exactly two pure strategies. We now know that there are no Nash equilibria of these types.

The same ideas work more generally for two-player constant-sum games, in which the two players' payoffs always add up to the same constant. In Tennis, for example, they always add up to 1.

5.6   The Ultimatum Minigame

This section deals with a simplified version of the Ultimatum Game of Section 1.7.

There are two players, Alice and Bob. The game organizer gives Alice $4. She can make Bob a fair offer of $2 or an unfair offer of $1. If Bob accepts Alice's offer, the $4 is divided accordingly between the two players. If Bob rejects the offer, the $4 goes back to the game organizer.

We shall assume that if Alice makes the fair offer, Bob will always accept it. Then the game tree is given by Figure 5.3.

Backward induction predicts that Alice will make an unfair offer, and Bob will accept it. This correspond to the prediction of backward induction in the Ultimatum Game.

Alice has two strategies: make a fair offer (f) or make an unfair offer (u). Bob also has two strategies: accept an unfair offer (a) or reject an unfair offer (r). Whichever strategy Bob adopts, he will accept a fair offer. The normal form of the game is given by the following matrix:

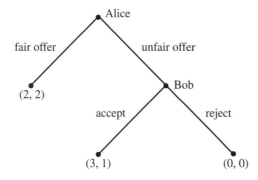

Figure 5.3. Ultimatum Minigame. Alice is Player 1, Bob is Player 2.

		Bob	
		a	r
Alice	f	$(2,2)$	$(2,2)$
	u	$(3,1)$	$(0,0)$

There are two pure strategy Nash equilibria: (u,a), a strict Nash equilibrium in which Alice makes an unfair offer and Bob accepts it; and (f,r), a Nash equilibrium that is not strict in which Alice makes a fair offer because Bob threatens to reject an unfair offer.

The second Nash equilibrium is not strict because, if Alice uses f, Bob's strategy a gives him the same payoff as his strategy r. In this situation we should look for Nash equilibria in which Alice uses one strategy f but Bob uses both strategies a and r.

Let's look at this Nash equilibrium (f,r) from the point of view of the Fundamental Theorem of Nash Equilibria (Theorem 5.2). The inactive strategies are Alice's strategy u and Bob's strategy a. In this case the Fundamental Theorem says that to have a Nash equilibrium, the following inequalities must hold:

$$\pi_1(f,r) \geqslant \pi_1(u,r) \quad \text{and} \quad \pi_2(f,r) \geqslant \pi_2(f,a).$$

Both inequalities are true, but the second is actually an equality.

Advice to remember when looking for mixed strategy Nash equilibria: Suppose a strategy profile σ^* satisfies the equality conditions for a Nash equilibrium, but for one or more inactive strategies s_i', the corresponding inequality condition is actually an equality. Then there may be Nash equilibria in which, in addition to the strategies that are active in σ^*, one or more of the strategies s_i' is active. Even in a two-player game, these Nash equilibria may be ones in which the players use different numbers of active strategies.

Following this advice, we examine strategy profiles $(f, qa + (1-q)r)$ with $0 < q < 1$. For such a strategy profile to be a Nash equilibrium, each of

Bob's pure strategies a and r should give Bob the same payoff against Alice's strategy f. Of course this is true: each gives Bob a payoff of 2. Thus every strategy profile of the form $(f, qa + (1 - q)r)$ with $0 < q < 1$ satisfies the equality conditions for a Nash equilibrium.

Finally we check the inequality condition: Alice's strategy f should give her at least as good a payoff against Bob's mixed strategy as does her unused strategy u, so we must have $\pi_1(f, qa + (1 - q)r) \geqslant \pi_1(u, qa + (1 - q)r)$. This inequality yields

$$2q + 2(1 - q) \geqslant 3q \quad \text{or} \quad q \leqslant \tfrac{2}{3}.$$

Thus, to induce Alice to make a fair offer, Bob does not have to threaten to definitely reject an unfair offer. It is enough to threaten to reject an unfair offer with probability $1 - q \geqslant \tfrac{1}{3}$.

5.7 Colonel Blotto vs. the People's Militia

There are two valuable towns. Col. Blotto has four regiments. The People's Militia has three regiments. Each decides how many regiments to send to each town.

If Col. Blotto sends m regiments to a town and the People's Militia sends n, Col. Blotto's payoff for that town is

$$
\begin{aligned}
1 + n & \quad \text{if } m > n, \\
0 & \quad \text{if } m = n, \\
-(1 + m) & \quad \text{if } m < n.
\end{aligned}
$$

Col. Blotto's total payoff is the sum of his payoffs for each town. The People's Militia's payoff is the opposite of Col. Blotto's.

We consider this to be a game in normal form. Col. Blotto (Player 1) has five strategies, which we denote 40, 31, 22, 13, and 04. Strategy 40 is to send four regiments to town 1 and none to town 2, and so forth for the other strategies. Similarly, the People's Militia (Player 2) has four strategies, which we denote 30, 21, 12, and 03.

We shall look for a mixed strategy Nash equilibrium (σ_1, σ_2), with $\sigma_1 = p_1 40 + p_2 31 + p_3 22 + p_4 13 + p_5 04$ and $\sigma_2 = q_1 30 + q_2 21 + q_3 12 + q_4 03$ (Table 5.1).

We should consider the following possibilities:

(1) Both players use a pure strategy.
(2) Both players use exactly two active strategies.
(3) Both players use exactly three active strategies.
(4) Both players use exactly four active strategies.

Table 5.1. Payoffs for Col. Blotto vs. the People's Militia

			People's Militia			
			q_1 30	q_2 21	q_3 12	q_4 03
	p_1	40	$(4, -4)$	$(2, -2)$	$(1, -1)$	$(0, 0)$
	p_2	31	$(1, -1)$	$(3, -3)$	$(0, 0)$	$(-1, 1)$
Col. Blotto	p_3	22	$(-2, 2)$	$(2, -2)$	$(2, -2)$	$(-2, 2)$
	p_4	13	$(-1, 1)$	$(0, 0)$	$(3, -3)$	$(1, -1)$
	p_5	04	$(0, 0)$	$(1, -1)$	$(2, -2)$	$(4, -4)$

In addition, we will look briefly at the following possibility:

(5) Both players use all their active strategies (five for Col. Blotto, four for the People's Militia).

5.7.1 Possibility 1. If both players use a pure strategy, we have a pure strategy Nash equilibrium. You can check using best response that there are none.

5.7.2 Possibility 2. We will not look at any possibilities in which both players use exactly two active strategies.

5.7.3 Possibility 3. Suppose both players use exactly three active strategies. There are 40 ways this can happen. (Col. Blotto has 10 ways to choose 3 of his 5 strategies; the People's Militia has 4 ways to choose 3 of their 4 strategies; $10 \times 4 = 40$.) We consider just two of these choices.

Choice 1. Suppose Col. Blotto uses only his 40, 31, and 22 strategies, and the People's Militia uses only its 30, 21, and 12 strategies. Thus we look for a Nash equilibrium (σ_1, σ_2), $\sigma_1 = p_1 40 + p_2 31 + p_3 22$, $\sigma_2 = q_1 30 + q_2 21 + q_3 12$.

Each of the People's Militia's pure strategies 30, 21, and 12 must yield the same payoff to the People's Militia against Col. Blotto's mixed strategy σ_1. These three payoffs are

$$\pi_2(\sigma_1, 30) = -4p_1 - p_2 + 2p_3,$$
$$\pi_2(\sigma_1, 21) = -2p_1 - 3p_2 - 2p_3,$$
$$\pi_2(\sigma_1, 12) = -p_1 - 2p_3.$$

We obtain three equations in three unknowns by using

$$\pi_2(\sigma_1, 30) = \pi_2(\sigma_1, 21) \quad \text{and} \quad \pi_2(\sigma_1, 30) = \pi_2(\sigma_1, 12),$$

together with $p_1 + p_2 + p_3 = 1$:

$$-2p_1 + 2p_2 + 4p_3 = 0,$$
$$-3p_1 - p_2 + 4p_3 = 0,$$
$$p_1 + p_2 + p_3 = 1.$$

The solution is $(p_1, p_2, p_3) = (\frac{3}{4}, -\frac{1}{4}, \frac{1}{2})$. Since p_2 is not strictly between 0 and 1, there is no Nash equilibrium of the desired type.

Choice 2. Suppose Col. Blotto uses only his 40, 22, and 04 strategies, and the People's Militia uses only its 30, 21, and 12 strategies. Thus we look for a Nash equilibrium (σ_1, σ_2), $\sigma_1 = p_1 40 + p_3 22 + p_5 04$, $\sigma_2 = q_1 30 + q_2 21 + q_3 12$.

Each of the People's Militia's pure strategies 30, 21, and 12 must yield the same payoff to the People's Militia against Col. Blotto's mixed strategy σ_1. These three payoffs are

$$\pi_2(\sigma_1, 30) = -4p_1 + 2p_3,$$
$$\pi_2(\sigma_1, 21) = -2p_1 - 2p_3 - p_5,$$
$$\pi_2(\sigma_1, 12) = -p_1 - 2p_3 - 2p_5.$$

We obtain three equations in three unknowns by using

$$\pi_2(\sigma_1, 30) = \pi_2(\sigma_1, 12) \quad \text{and} \quad \pi_2(\sigma_1, 21) = \pi_2(\sigma_1, 12),$$

together with $p_1 + p_3 + p_5 = 1$:

$$-3p_1 + 4p_3 + 2p_5 = 0,$$
$$-p_1 + p_5 = 0,$$
$$p_1 + p_3 + p_5 = 1.$$

The solution is $(p_1, p_3, p_5) = (\frac{4}{9}, \frac{1}{9}, \frac{4}{9})$.

Each of Col. Blotto's pure strategies 40, 22, and 04 must yield the same payoff to Col. Blotto against the People's Militia's mixed strategy σ_2. These three payoffs are

$$\pi_1(40, \sigma_2) = 4q_1 + 2q_2 + q_3,$$
$$\pi_1(22, \sigma_2) = -2q_1 + 2q_2 + 2q_3,$$
$$\pi_1(04, \sigma_2) = q_2 + 2q_3.$$

We obtain three equations in three unknowns by using

$$\pi_1(40, \sigma_2) = \pi_1(04, \sigma_2) \quad \text{and} \quad \pi_1(22, \sigma_2) = \pi_1(04, \sigma_2),$$

together with $q_1 + q_2 + q_3 = 1$:

$$4q_1 + q_2 - q_3 = 0,$$
$$-2q_1 + q_2 = 0,$$
$$q_1 + q_2 + q_3 = 1.$$

The solution is $(q_1, q_2, q_3) = (\frac{1}{9}, \frac{2}{9}, \frac{2}{3})$.

These calculations rule out the existence of Nash equilibria in which the People's Militia's active strategies are 30, 21, and 12, and Col. Blotto's active strategies are one or two of his strategies 40, 22, and 04. They also rule out the existence of Nash equilibria in which Col. Blotto's active strategies are 40, 22, and 04, and the People's Militia's active strategies are one or two of their strategies 30, 21, and 12.

We have seen that (σ_1, σ_2) with $\sigma_1 = \frac{4}{9}40 + \frac{1}{9}22 + \frac{4}{9}04$ and $\sigma_2 = \frac{1}{9}30 + \frac{2}{9}21 + \frac{2}{3}12$ satisfies the equality conditions for a Nash equilibrium. We now check the inequality conditions.

For Col. Blotto:

$$\pi_1(40, \sigma_2) = \pi_1(22, \sigma_2) = \pi_1(04, \sigma_2) = \frac{14}{9},$$
$$\pi_1(31, \sigma_2) = \frac{1}{9}(1) + \frac{2}{9}(3) + \frac{2}{3}(0) = \frac{7}{9},$$
$$\pi_1(13, \sigma_2) = \frac{1}{9}(-1) + \frac{2}{9}(0) + \frac{2}{3}(3) = \frac{17}{9}$$

For the People's Militia:

$$\pi_2(\sigma_1, 30) = \pi_2(\sigma_1, 21) = \pi_2(\sigma_1, 12) = -\frac{14}{9},$$
$$\pi_2(\sigma_1, 03) = -\frac{14}{9}$$

Since $\pi_1(13, \sigma_2) > \frac{14}{9}$, the inequality conditions are not satisfied; (σ_1, σ_2) is not a Nash equilibrium.

Notice, however, that $\pi_2(\sigma_1, 03) = -\frac{14}{9}$, that is, Player 2's strategy 03 does just as well against σ_1 as the strategies that are active in σ_2. *As discussed in Section 5.6, when this happens, it is possible that there is a mixed strategy Nash equilibrium in which the two players use different numbers of active strategies.* In this case, we must check the possibility that there is a Nash equilibrium in which Col. Blotto's active strategies are 40, 22, and 04, and the People's Militia's strategies include 03 in addition to 30, 21, and 12.

Addendum to Choice 2. Thus we suppose Col. Blotto's active strategies are 40, 22, and 04, and all of the People's Militia's strategies are active. In other words, we look for a Nash equilibrium $\sigma = (\sigma_1, \sigma_2)$, $\sigma_1 = p_1 40 + p_3 22 + p_5 04$, $\sigma_2 = q_1 30 + q_2 21 + q_3 12 + q_4 03$.

Each of the People's Militia's pure strategies must yield the same payoff to the People's Militia against Col. Blotto's mixed strategy σ_1. We obtain four equations in three unknowns by using $\pi_2(\sigma_1, 30) = \pi_2(\sigma_1, 03)$, $\pi_2(\sigma_1, 21) = \pi_2(\sigma_1, 03)$, and $\pi_2(\sigma_1, 12) = \pi_2(\sigma_1, 03)$, together with $p_1 + p_3 + p_5 = 1$. Usually, if there are more equations than unknowns, there are no solutions. In this case, however, there is a solution: $(p_1, p_3, p_5) = (\frac{4}{9}, \frac{1}{9}, \frac{4}{9})$; that is, the same solution we had before allowing the People's Militia to use its strategy 03. A little thought indicates that this is what will always happen.

Each of Col. Blotto's pure strategies 40, 22, and 04 must yield the same payoff to Col. Blotto against the People's Militia's mixed strategy σ_2. These three payoffs are

$$\pi_1(40, \sigma_2) = 4q_1 + 2q_2 + q_3,$$
$$\pi_1(22, \sigma_2) = -2q_1 + 2q_2 + 2q_3 - 2q_4,$$
$$\pi_1(04, \sigma_2) = q_2 + 2q_3 + 4q_4.$$

We obtain three equations in four unknowns by using $\pi_1(40, \sigma_2) = \pi_1(04, \sigma_2)$ and $\pi_1(22, \sigma_2) = \pi_1(04, \sigma_2)$, together with $q_1 + q_2 + q_3 + q_4 = 1$:

$$4q_1 + q_2 - q_3 - 4q_4 = 0,$$
$$-2q_1 + q_2 - 6q_4 = 0,$$
$$q_1 + q_2 + q_3 + q_4 = 1.$$

As usually happens with fewer equations than unknowns, there are many solutions. One way to list them all is as follows:

$$q_1 = \tfrac{1}{9} - q_4, \quad q_2 = \tfrac{2}{9} + 4q_4, \quad q_3 = \tfrac{2}{3} - 4q_4, \quad q_4 \text{ arbitrary.}$$

To keep all the q_is strictly between 0 and 1, we must restrict q_4 to the interval $0 < q_4 < \tfrac{1}{9}$.

Thus, if $\sigma_1 = \tfrac{4}{9}40 + \tfrac{1}{9}22 + \tfrac{4}{9}04$ and

$$\sigma_2 = (\tfrac{1}{9} - q_4)30 + (\tfrac{2}{9} + 4q_4)21 + (\tfrac{2}{3} - 4q_4)12 + q_4 03, \quad 0 < q_4 < \tfrac{1}{9},$$

then (σ_1, σ_2) satisfies the equality conditions for a Nash equilibrium. We now consider the inequality conditions when $0 < q_4 < \tfrac{1}{9}$, so that all of the People's Militia's strategies are active. Then for the People's Militia, there is no inequality constraint to check. For Col. Blotto:

$$\pi_1(40, \sigma_2) = \pi_1(22, \sigma_2) = \pi_1(04, \sigma_2) = \tfrac{14}{9},$$
$$\pi_1(31, \sigma_2) = (\tfrac{1}{9} - q_4)(1) + (\tfrac{2}{9} + 4q_4)(3) + (\tfrac{2}{3} - 4q_4)(0) + q_4(-1)$$
$$= 10q_4 + \tfrac{7}{9},$$
$$\pi_1(13, \sigma_2) = (\tfrac{1}{9} - q_4)(-1) + (\tfrac{2}{9} + 4q_4)(0) + (\tfrac{2}{3} - 4q_4)(3) + q_4(1)$$
$$= -10q_4 + \tfrac{17}{9}.$$

To satisfy the inequality constraints for a Nash equilibrium, we need

$$10q_4 + \tfrac{7}{9} \leqslant \tfrac{14}{9} \quad \text{and} \quad -10q_4 + \tfrac{17}{9} \leqslant \tfrac{14}{9}.$$

These inequality conditions are satisfied for $\tfrac{1}{30} \leqslant q_4 \leqslant \tfrac{7}{90}$.

We have thus found a one-parameter family of Nash equilibria (σ_1, σ_2):
$\sigma_1 = \tfrac{4}{9}40 + \tfrac{1}{9}22 + \tfrac{4}{9}04$ and

$$\sigma_2 = (\tfrac{1}{9} - q_4)30 + (\tfrac{2}{9} + 4q_4)21 + (\tfrac{2}{3} - 4q_4)12 + q_4 03, \quad \tfrac{1}{30} \leqslant q_4 \leqslant \tfrac{7}{90},$$

The most attractive of these Nash equilibria occurs for q_4 at the midpoint of its allowed interval of values: $(q_1, q_2, q_3, q_4) = (\tfrac{1}{18}, \tfrac{4}{9}, \tfrac{4}{9}, \tfrac{1}{18})$. At this Nash equilibrium, the People's Militia uses its 30 and 03 strategies equally, and also uses its 21 and 12 strategies equally.

We shall discuss this "symmetric" Nash equilibrium further in Section 7.4.

5.7.4 Possibility 4.

We will not look at any possibilities in which both players use four active strategies.

5.7.5 Possibility 5.

Suppose Col. Blotto uses all five of his pure strategies. Then at a Nash equilibrium, each of Col. Blotto's five pure strategies gives the same payoff to him against the People's Militia's mixed strategy σ_2. Therefore we have the following system of 5 equations in the 4 unknowns q_1, q_2, q_3, q_4:

$$\pi_1(40, \sigma_2) = \pi_1(04, \sigma_2),$$
$$\pi_1(31, \sigma_2) = \pi_1(04, \sigma_2),$$
$$\pi_1(22, \sigma_2) = \pi_1(04, \sigma_2),$$
$$\pi_1(13, \sigma_2) = \pi_1(04, \sigma_2),$$
$$q_1 + q_2 + q_3 + q_4 = 1.$$

Typically, when there are more equations than unknowns, there is no solution. One can check that that is the case here.

The game we have discussed is one of a class called Colonel Blotto games. They differ in the number of towns and in the number of regiments available to Col. Blotto and his opponent. There is a Wikipedia page devoted to these games: http://en.wikipedia.org/wiki/Colonel_Blotto. There you will learn that it has been argued that U.S. presidential campaigns should be thought of as Colonel Blotto games, in which the candidates must allocate their resources among the different states.

5.8 Water Pollution 3

In the game of Water Pollution (Section 3.3), we have already considered pure strategy Nash equilibria. Now we consider mixed strategy Nash equilibria in which all three players use completely mixed strategies. Let g and b denote the strategies purify and pollute, respectively. Then we search for a mixed strategy Nash equilibrium $(\sigma_1, \sigma_2, \sigma_3) = (xg + (1-x)b, yg + (1-y)b, zg + (1-z)b)$. Since the numbers x, y, and z determine the player's strategies, we shall think of the payoff functions π_i as functions of (x, y, z). Table 5.2 helps keep track of the notation.

Table 5.2. Payoff matrices for Water Pollution game

Firm 3 z g

			Firm 2 y g	Firm 2 $1-y$ b
Firm 1	x	g	$(-1,-1,-1)$	$(-1,0,-1)$
	$1-x$	b	$(0,-1,-1)$	$(-3,-3,-4)$

Firm 3 $1-z$ b

			Firm 2 y g	Firm 2 $1-y$ b
Firm 1	x	g	$(-1,-1,0)$	$(-4,-3,-3)$
	$1-x$	b	$(-3,-4,-3)$	$(-3,-3,-3)$

The criteria for a mixed strategy Nash equilibrium in which all three players have two active strategies (i.e., $0 < x < 1$, $0 < y < 1$, $0 < z < 1$) are:

$$\pi_1(1, y, z) = \pi_1(0, y, z),$$
$$\pi_2(x, 1, z) = \pi_2(x, 0, z),$$
$$\pi_3(x, y, 1) = \pi_3(x, y, 0).$$

The first equation, for example, says that if Players 2 and 3 use the mixed strategies $yg + (1-y)b$ and $zg + (1-z)b$, respectively, then the payoff to Player 1 if he uses g must equal the payoff to him if he uses b.

The first equation, written out, is

$$-yz - (1-y)z - y(1-z) - 4(1-y)(1-z)$$
$$= 0yz - 3(1-y)z - 3y(1-z) - 3(1-y)(1-z),$$

which simplifies to

$$1 - 3y - 3z + 6yz = 0. \tag{5.12}$$

The other two equations, after simplification, are

$$1 - 3x - 3z + 6xz = 0 \quad \text{and} \quad 1 - 3x - 3y + 6xy = 0. \tag{5.13}$$

A straightforward way to solve this system of three equations in the unknowns x, y, and z is to begin by solving the last two equations for z in terms of x and for y in terms of x. We obtain

$$z = \frac{1 - 3x}{3 - 6x}, \qquad y = \frac{1 - 3x}{3 - 6x}. \tag{5.14}$$

Therefore $z = y$. Similarly, if we solve the first equation for y in terms of z and the second for x in terms of z, we find that $y = x$. Hence $x = y = z$. Now we set $z = y$ in the first equation, which yields $6y^2 - 6y + 1 = 0$. The quadratic formula then gives $y = \frac{1}{2} \pm \frac{1}{6}\sqrt{3}$. We have therefore found two mixed strategy Nash equilibria: $x = y = z = \frac{1}{2} + \frac{1}{6}\sqrt{3}$ and $x = y = z = \frac{1}{2} - \frac{1}{6}\sqrt{3}$.

5.9 Equivalent games

Assigning payoffs in games is a tricky business. Are there some aspects of the process that we can safely ignore?

When you do backward induction on a game in extensive form with complete information, or when you eliminate dominated strategies or look for pure strategy Nash equilibria in a game in normal form, all you need to do is compare one payoff of Player i to another payoff of the same player. Therefore any reassignment of Player i's payoffs that preserves their order will not affect the outcome of the analysis.

The situation is different when you work with a game in extensive form with incomplete information, or when you look for mixed strategy Nash equilibria in a game in normal form. Suppose Player i's payoffs in the different situations are $v_1, \ldots, v_k$. The safe reassignments of payoffs are *affine linear*: choose numbers $a > 0$ and b, and replace each v_j by $av_j + b$. You may use a different a and b for each player.

We will not give a proof of this, but it is intuitively obvious. Suppose, for example, that Player i's payoffs are in dollars, and we replace each v_i by $100v_i + 200$. This can be interpreted as calculating Player i's payoffs in cents instead of dollars, and giving Player i two dollars each time she plays the game, independent of how she does. Neither of these changes should affect Player i's choice of strategies in the game.

Example. The payoff matrix for a prisoner's dilemma with two players and two strategies takes the form

		Player 2	
		c	d
Player 1	c	(r_1, r_2)	(s_1, t_2)
	d	(t_1, s_2)	(p_1, p_2)

with $s_i < p_i < r_i < t_i$ for each i. This is a standard notation. The strategies are cooperate (c) and defect (d). The letter s stands for sucker's payoff, p for punishment (the lousy payoff when both players defect), t for temptation, and r for reward (the good payoff when both cooperate). Note that d strictly dominates c for both players, but the strategy profile (c, c) yields better payoffs to both than the strategy profile (d, d).

Let us try to replace Player 1's payoffs v by new payoffs $av + b$ so that r_1 becomes 1 and p_1 becomes 0. Thus we want to choose a and b so that

$$ar_1 + b = 1,$$
$$ap_1 + b = 0.$$

We find that $a = 1/(r_1 - p_1)$ and $b = -p_1/(r_1 - p_1)$. Similarly, we replace Player 2's payoffs v by new payoffs $av + b$ with $a = 1/(r_2 - p_2)$ and $b = -p_2/(r_2 - p_2)$. We obtain the new payoff matrix

		Player 2	
		c	d
Player 1	c	$(1, 1)$	(e_1, f_2)
	d	(f_1, e_2)	$(0, 0)$

with

$$e_i = \frac{s_i - p_i}{r_i - p_i} < 0 \quad \text{and} \quad f_i = \frac{t_i - p_i}{r_i - p_i} > 1.$$

You might find the second payoff matrix easier to think about than the first. Also, if one were interested in doing a complete analysis of these games, the second payoff matrix would be preferable to the first, because it has four parameters instead of eight.

5.10 Software for computing Nash equilibria

When the game is large, computing Nash equilibria can become too difficult, or at least too tedious, to do by hand.

To find a Nash equilibrium, according to the Fundamental Theorem of Nash Equilibria, one must do the following:

(1) Determine which strategies will be active.
(2) Find a solution to a set of equations with all variables nonnegative.
(3) If not all strategies are active, check some further inequalities.

There are two nice features of this problem: in step 2, the inequalities are linear ($p_{ij} \geqslant 0$); and step 3, since a solution to the equalities has already been found, is just arithmetic.

For an n-player game, the equations in step 2 are polynomial of degree $n-1$. For example, in the three-player game Water Pollution, we found second-degree polynomials in Section 5.8.

The easiest case is therefore two-player games, which give rise to linear equations. In fact, for two-player games, there are alternate approaches to computing Nash equilibria that do not directly use the Fudamental Theorem. For a two-player zero-sum game, the problem of finding a Nash equilibrium can be converted into a *linear programming problem*, for which many solution methods are known. For general two-player games, one can use the *Lemke-Howson algorithm*, which is related to the simplex algorithm of linear programming.

For n-player games, one first uses a *heuristic* (an experience-based algorithm that is not guaranteed to work) to choose a likely set of active strategies. Then one can use computer algebra or a numerical method to find solutions to the resulting system of polynomial equations and linear equalities. Finally, one checks whether the inequalities in the Fundamental Theorem are satisfied.

The free software Gambit, available at http://www.gambit-project.org, incorporates a variety of methods for computing Nash equilibria numerically. Gambit can also be used to eliminate dominated strategies in normal-form games, and to perform backward induction in extensive-form games.

5.11 Critique of Nash equilibrium

Let's consider the game of Tennis (Section 5.2). Suppose you must serve *once*, and you know that your opponent is using her mixed strategy from the Nash equilibrium: she expects a serve to her backhand 70% of the time, and a serve to her forehand 30% of the time. Should you use your own mixed strategy from the Nash equilibrium, namely, serve with probability 60% to her forehand and with probability 40% to her backhand? The answer is that it doesn't matter: according to Theorem 5.3, either of your pure strategies, or any mixture of them, gives you the same expected payoff.

But what if you must serve many times, as is actually the case in a game of tennis? If you do not respond to her 70-30 expectations with the corresponding 60-40 strategy, then her 70-30 strategy is not her best response to your strategy. If she realizes what you are doing, she can change strategies and improve her payoff.

More generally, if the two players are not in a Nash equilibrium, one player might eventually realize that she can improve her own payoff by changing her strategy. If the new strategy profile is not a Nash equilibrium, again one of the players might eventually realize that she can improve her payoff by changing her strategy. Thus one would not expect the players to stay with a pair of strategies that is not a Nash equilibrium. However, these repeated adjustments need not lead the players toward a Nash equilibrium. In the game of Tennis, if one player does not use her Nash equilibrium mixed strategy, the other player's best response is always a pure strategy. Thus if the players repeatedly adjust to each other using best responses, they will never arrive at a Nash equilibrium. In Chapter 10 we look at a different adjustment process that makes small adjustments. However, it does not necessarily need to result in a Nash equilibrium either.

One view is that rational analysis of a game by the players will lead them to play a Nash equilibrium. Another view is that when a game is played many times, trial and error leads to a Nash equilibrium without the necessity for rational analysis. At present there do not exist either theoretical arguments or experimental results to fully justify either view, although there is some evidence that professional athletes do use mixed strategies close to the Nash equilibrium ones in situations like that of Tennis.

We have seen that game theory provides models of interaction that can be used to understand and deal with recreational games; ordinary human interactions; animal behavior; and issues of politics, economics, business, and war. This list is not exhaustive. The role played by the Nash equilibrium may well depend on the context in which it arises.

5.12 Problems

5.12.1 Courtship among Birds. In many bird species, males are faithful or philanderers, females are coy or loose. Coy females insist on a long courtship before copulating, while loose females do not. Faithful males tolerate a long courtship and help rear their young, while philanderers do not wait and do not help. Suppose v is the value of having offspring to either a male or a female, $2r > 0$ is the total cost of rearing offspring, and $w > 0$ is the cost of prolonged courtship to both male and female. We assume $v > r + w$. This means that a long courtship followed by sharing the costs of raising offspring is worthwhile to both male and female birds. The normal form of this game is:

		Female	
		coy	**loose**
Male	**faithful**	$(v - r - w, v - r - w)$	$(v - r, v - r)$
	philanderer	$(0, 0)$	$(v, v - 2r)$

(1) If $v > 2r$, find a pure strategy Nash equilibrium.

(2) If $v < 2r$, show that there is no pure strategy Nash equilibrium, and find a mixed strategy Nash equilibrium.

5.12.2 The Travails of Boss Gorilla 3.

In Section 4.5, where Boss Gorilla had to deal with a Visiting Gorilla whose unknown type was Tough or Weak, we found a pure strategy Nash equilibrium under the assumption that Visiting Gorilla was Tough with probability $p > \frac{1}{3}$. Now suppose that $p < \frac{1}{3}$. To be specific, suppose that $p = \frac{1}{6}$. Then the payoff matrix for the two-player game in Section 4.5, with Visiting Gorilla's strictly dominated strategies lc and ll removed, is:

		Visiting Gorilla	
		cc	**cl**
Boss Gorilla	**f**	$(4, 0)$	$(4\frac{5}{6}, \frac{5}{6})$
	a	$(3, 3)$	$(5\frac{1}{2}, \frac{1}{2})$

(1) Use best response to show that there is no pure strategy Nash equilibrium.

(2) Find a mixed strategy Nash equilibrium.

(3) Interpret your answer to part (2): in the Nash equilibrium, what should each gorilla (Boss, Tough, and Weak) do?

5.12.3 War and Peace.

Two players each have one unit of Good Stuff. They each have two strategies, remain peaceful (p) or attack (a). If both remain peaceful, each gets to consume his own stuff. If one attacks and the other remains peaceful, the attacker takes the other's stuff. If both attack, both incur a loss of $\ell > 0$. The normal form of this game is:

		Player 2	
		p	**a**
Player 1	**p**	$(1, 1)$	$(0, 2)$
	a	$(2, 0)$	$(-\ell, -\ell)$

(1) Use best response to find the pure strategy Nash equilibria. (There are two.)

(2) Find a mixed strategy Nash equilibrium in which neither player uses a pure strategy.

(3) Show that for the mixed strategy Nash equilibrium, the payoff to each player increases when the loss from conflict ℓ increases. Interpret this result.

5.12.4 Rock-Paper-Scissors 1. In the well-known game of Rock-Paper-Scissors, two players simultaneously form an outstretched hand into one of three shapes: rock (closed fist), paper (open hand), or scissors (a fist with index and middle finger stuck out). Paper beats rock, scissors beats paper, and rock beats scissors. If we assign a payoff of 1 to the winner, -1 to the loser, and 0 to both players in case of a tie, we get the following payoff matrix:

		Player 2		
		r	p	s
	r	$(0,0)$	$(-1,1)$	$(1,-1)$
Player 1	p	$(1,-1)$	$(0,0)$	$(-1,1)$
	s	$(-1,1)$	$(1,-1)$	$(0,0)$

(1) Use best response to show that there are no pure strategy Nash equilibria.
(2) Find a mixed strategy Nash equilibrium in which both players use all three of their pure strategies.
(3) Show that in the mixed strategy Nash equilibria found in part (2), the expected payoff to each player is 0.
(4) Use Theorem 5.6 to explain why the Nash equilibrium found in part (2) is the only Nash equilibrium of Rock-Paper-Scissors. (Take a look at how Theorem 5.6 was used at the end of Section 5.5 to rule out additional Nash equilibria in the game of One-card Two-round Poker.)

For more information about Rock-Paper-Scissors, see the Wikipedia article http://en.wikipedia.org/wiki/Rock-paper-scissors.

5.12.5 Product Development. Two companies are racing to be first to develop a product. Company 1 can invest 0, 1, 2, 3, or 4 million dollars in this effort. Company 2, which is a little bigger, can invest 0, 1, 2, 3, 4, or 5 million dollars. If one company invests more, it will win the race and gain 10 million dollars by being first to market. If the companies invest the same amount, there is no gain to either. Total payoff to each company is the amount it did *not* invest, which it retains, plus 10 million if it is first to market. The normal form of this game is:

	0	1	2	3	4	5
0	(4, 5)	(4, 14)	(4, 13)	(4, 12)	(4, 11)	(4, 10)
1	(13, 5)	(3, 4)	(3, 13)	(3, 12)	(3, 11)	(3, 10)
2	(12, 5)	(12, 4)	(2, 3)	(2, 12)	(2, 11)	(2, 10)
3	(11, 5)	(11, 4)	(11, 3)	(1, 2)	(1, 11)	(1, 10)
4	(10, 5)	(10, 4)	(10, 3)	(10, 2)	(0, 1)	(0, 10)

(1) Show that after eliminating weakly dominated strategies, each player has three remaining strategies.

(2) Show that the reduced 3×3 game has no pure strategy Nash equilibria.

(3) Find a mixed strategy Nash equilibrium of the reduced game in which each company uses all three of its remaining strategies.

(4) Show that in this mixed strategy Nash equilibria, the expected payoffs are 4 to company 1 and 10 to company 2.

5.12.6 Tit for Tat 2. Consider Al's and Bob's toy stores, as described in Problem 3.12.6. Recall that each had three strategies:

- h: Always charge high prices.
- l: Always charge low prices.
- t: Tit for tat. Charge high prices the first week. The next week, do whatever the other store did the previous week.

The matrix of payoffs if each store follows its strategy for two weeks is:

		Bob		
		h	l	t
	h	(10,10)	(0,12)	(10,10)
Al	l	(12,0)	(6,6)	(9,3)
	t	(10,10)	(3,9)	(10,10)

We shall look for a mixed strategy Nash equilibrium (σ, τ) in which each store uses a combination of its strategies h and t: $\sigma = ph + (1 - p)t$, $\tau = qh + (1 - q)t$, $0 < p < 1$, $0 < q < 1$.

(1) The equality conditions for a Nash equilibrium of this type are $\pi_1(h, \tau) = \pi_1(t, \tau)$ and $\pi_2(\sigma, h) = \pi_2(\sigma, t)$. Show that these equations place no restriction at all on (p, q). Hence every strategy profile (σ, τ) defined above satisfies the equality conditions for a Nash equilibrium.

(2) Now we consider the inequality condition for Player 1. Let $\tau = qh + (1 - q)t$, $0 < q < 1$. We have $\pi_1(h, \tau) = \pi_1(t, \tau) = 10$. Therefore the inequality condition is $\pi_1(l, \tau) \leqslant 10$. Find the restriction that this places on q.

5.12.7 Smallville Bar. The town of Smallville has three residents. At night, each has two choices: watch TV (*t*) or walk to the bar (*b*). The energy cost of watching TV is 0, and the utility is also 0. The energy cost of walking to the bar is 1; the utility is 0 if no one else is at the bar, 2 if one other resident is at the bar, and 1 if both other residents are at the bar. (The residents of Smallville are sociable, but not too sociable.) The payoffs are shown in Table 5.3.

Table 5.3. Payoff matrices for Smallville Bar game

Resident 3 uses strategy *t*

		Resident 2 *t*	Resident 2 *b*
Resident 1	*t*	(0,0,0)	(0,−1,0)
	b	(−1,0,0)	(1,1,0)

Resident 3 uses strategy *b*

		Resident 2 *t*	Resident 2 *b*
Resident 1	*t*	(0,0,−1)	(0,1,1)
	b	(1,0,1)	(0,0,0)

(1) Use best response to find the pure strategy Nash equilibria.
(2) Suppose Resident 1 uses the mixed strategy $xt + (1 - x)b$, Resident 2 uses the mixed strategy $yt + (1 - y)b$, and Resident 3 uses the mixed strategy $zt + (1 - z)b$. As in Section 5.8, add the letters x, $1 - x$, and so forth to the matrices above in the appropriate places to help keep track of the notation.
(3) Find a Nash equilibrium in which no resident uses a pure strategy. (Answer: $(x, y, z) = (\frac{2}{3}, \frac{2}{3}, \frac{2}{3})$.)
(4) Find the expected payoff of Resident 1 at the Nash equilibrium. (It is 0.)

If the residents use the Nash equilibrium strategies of part (4), then their expected utility is the same as the utility of watching TV, even though they go to the bar two-thirds of the time.

This problem is loosely based on a game theory classic, the El Farol Bar problem (http://en.wikipedia.org/wiki/El_Farol_Bar_problem). The El Farol Bar is a real bar in Santa Fe, New Mexico.

5.12.8 Morning or Evening? Three firms (Firms 1, 2, and 3) can advertise on TV in either the morning (m) or the evening (e). If more than one firm advertises at the same time, their profits are 0. If exactly one firm advertises in the morning, its profit is 1. If exactly one firm advertises in the evening, its profit is 2.

(1) Each firm has two strategies, m and e. Give the payoffs to each triple of pure strategies. You should organize your answer by giving two 2×2 matrices; see Section 3.10.

(2) Use best response to find the pure strategy Nash equilibria. (Answer: Every triple of pure strategies except (m, m, m) and (e, e, e).)

(3) Suppose Firm 1 uses the mixed strategy $xm + (1 - x)e$, Firm 2 uses the mixed strategy $ym + (1 - y)e$, and Firm 3 uses the mixed strategy $zm + (1 - z)e$. Show that the payoff functions are

$$\pi_1(x, y, z) = 2(1 - x)yz + x(1 - y)(1 - z),$$
$$\pi_2(x, y, z) = 2(1 - y)xz + y(1 - x)(1 - z),$$
$$\pi_3(x, y, z) = 2(1 - z)xy + z(1 - x)(1 - y).$$

(4) Suppose one player uses the pure strategy m and one uses the pure strategy e. Show that any mix of strategies by the third player yields a Nash equilibrium. (For example, for any z with $0 \leqslant z \leqslant 1$, $(m, e, zm + (1 - z)e)$ is a Nash equilibrium.)

(5) Show that there is no Nash equilibrium in which exactly one firm uses a pure strategy. (It is enough to show that there is no Nash equilibrium in which Firm 1 uses the pure strategy m or e, and Firms 2 and 3 use mixed strategies with $0 < y < 1$ and $0 < z < 1$.)

(6) Find a Nash equilibrium in which no firm uses a pure strategy.

5.12.9 Water Pollution 4. In the game of Water Pollution (Section 5.8), find a Nash equilibrium (σ_1, σ_2, g) in which

- Firm 1 uses the mixed strategy $\sigma_1 = xg + (1 - x)b$ with $0 < x < 1$.
- Firm 2 uses the mixed strategy $\sigma_2 = yg + (1 - y)b$ with $0 < y < 1$.
- Firm 3 uses the pure strategy g.

Suggestion: As usual you can use the Fundamental Theorem of Nash Equilibria. Applied to Firm 1 or Firm 2, the Fundamental Theorem gives an equation (an equality). Applied to Firm 3 it gives an inequality to check after you find x and y.

5.12.10 Guess the Number. Bob (Player 1) picks a number from 1 to 3. Alice (Player 2) tries to guess the number. Bob responds truthfully by saying "high," "low," or "correct." The game continues until Alice guesses correctly. Bob wins from Alice a number of dollars equal to the number of guesses that Alice took.

The game is determined by the first two rounds. Player 1 (Bob) has three strategies: pick 1, pick 2, pick 3. Player 2 (Alice) has five reasonable strategies:

- 12: Guess 1. If told it is low, guess 2.
- 13: Guess 1. If told it is low, guess 3.
- 2: Guess 2. If told it is high, guess 1. If told it is low, guess 3.
- 31: Guess 3. If told it is high, guess 1.
- 32: Guess 3. If told it is high, guess 2.

(1) Construct the payoff matrix. You should get the following answer:

	12	13	2	31	32
1	$(1,-1)$	$(1,-1)$	$(2,-2)$	$(2,-2)$	$(3,-3)$
2	$(2,-2)$	$(3,-3)$	$(1,-1)$	$(3,-3)$	$(2,-2)$
3	$(3,-3)$	$(2,-2)$	$(2,-2)$	$(1,-1)$	$(1,-1)$

(2) Use best response to find the pure strategy Nash equilibria. (There are none.)

(3) To look for mixed strategy Nash equilibria, let $\sigma_1 = (p_1, p_2, p_3)$ be a mixed strategy for Player 1, and let $\sigma_2 = (q_1, q_2, q_3, q_4, q_5)$ be a mixed strategy for Player 2. Find a Nash equilibrium in which all Player 1's strategies are active, and only Player 2's second, third, and fourth strategies are active.

(4) Determine whether there is a Nash equilibrium in which all Player 1's strategies are active, and only Player 2's first, third, and fifth strategies are active.

Chapter 6

More about games in extensive form with complete information

In this chapter we investigate some further aspects of games in extensive form with complete information.

First, we saw in Section 3.9 that when one converts a game in extensive form to normal form, new Nash equilibria can appear that do not correspond to the Nash equilibrium one finds by backward induction. The latter is special in that it gives a strategy one can sensibly use at any point in the game tree. In this chapter we give a name—subgame perfect Nash equilibria—and definition to these special Nash equilibria of a game in extensive form. The definition is general enough that it will help us deal with such games as the one in Figure 1.6, where the backward induction procedure we have used so far fails.

Second, we look at some infinite horizon games. Since infinite horizon games potentially last forever, they cannot be fully analyzed by backward induction. The notion of a subgame perfect Nash equilibrium will help us investigate these games.

Especially important are repeated games in which a prisoner's dilemma occurs over and over. In this context one can see why it might be advantageous to use the strictly dominated, cooperative strategy, if it induces cooperation from one's opponent in the future.

Infinite horizon games have more practical applications than one might think. For example, a game that is repeated an unknown number of times can often be thought of as an infinite horizon game.

Finally, we discuss two striking examples of games in extensive form with complete information, the Samaritan's Dilemma (Section 6.10) and the Rotten Kid Theorem (Section 6.11), that come from the work of Nobel Prize-winning economists. These two games were too complicated to include in Chapter 1, but in essence they are just applications of backward induction.

6.1 Subgame perfect Nash equilibria

Consider a game G in extensive form. A node c' is called a *successor* of a node c if there is a path in the game tree that starts at c and ends at c'.

Let h be a node that is not terminal and has no other nodes in its information set. Assume:

- If a node c is a successor of h, then every node in the information set of c is also a successor of h.

In this situation it makes sense to talk about the *subgame H* of G whose root is h. H consists of the node h and all its successors, connected by the same moves that connected them in G, and partitioned into the same information sets as in G. The players and the payoffs at the terminal vertices are also the same as in G.

If G is a game with complete information, then any nonterminal node of G is the root of a subgame of G.

Recall from Section 1.2 that in a game in extensive form, each node is at the end of a unique path from the root node. This is usually interpreted to mean that players remember the past. Thus you should think of a subgame as including the memory of what happened before the subgame began.

Let s_i be one of Player i's strategies in the game G. Recall that s_i is just a plan for what move to make at every node labeled i in the game G. So of course s_i includes a plan for what move to make at every node labeled i in the subgame H. Thus s_i contains within it a strategy that Player i can use in the subgame H. We call this strategy the *restriction* of s_i to the subgame H, and label it s_{iH}.

Suppose the game G has n players, and $(s_1,\ldots,s_n)$ is a Nash equilibrium for G. It is called a *subgame perfect Nash equilibrium* if, for every subgame H of G, $(s_{1H},\ldots,s_{nH})$ is a Nash equilibrium for H.

6.2 Big Monkey and Little Monkey 6

Recall the game of Big Monkey and Little Monkey from Section 1.5, with Big Monkey going first (Figure 6.1).

Recall that Big Monkey has two possible strategies in this game, and Little Monkey has four. When we find the payoffs for each choice of strategies, we get a game in normal form:

		Little Monkey			
		ww	wc	cw	cc
Big Monkey	w	$(0,0)$	$(0,0)$	$(9,1)$	$(9,1)$
	c	$(4,4)$	$(5,3)$	$(4,4)$	$(5,3)$

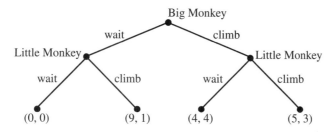

Figure 6.1. Big Monkey and Little Monkey.

We have seen (see Section 3.9) that there are three Nash equilibria in this game: (w, cw), (c, ww), and (w, cc). The equilibrium (w, cw), which is the one found by backward induction, is subgame perfect. The other two are not.

The equilibrium (c, ww) is not subgame perfect because, in the subgame that begins after Big Monkey waits, Little Monkey would wait. By switching to climb, Little Monkey would achieve a better payoff. Little Monkey's plan to wait if Big Monkey waits is what we called in Section 1.6 a *threat*: it would hurt Big Monkey, at the cost of hurting Little Monkey

The equilibrium (w, cc) is not subgame perfect because, in the subgame that begins after Big Monkey climbs, Little Monkey would climb. By switching to wait, Little Monkey would achieve a better payoff. Little Monkey's plan to climb if Big Monkey climbs is what we called in Section 1.6 a *promise*: it would help Big Monkey, at the cost of hurting Little Monkey. As we saw there, in this particular game, the promise does not affect Big Monkey's behavior. Little Monkey is promising that if Big Monkey climbs, he will get a payoff of 5, rather than the payoff of 4 he would normally expect. Big Monkey ignores this promise because by waiting, he gets an even bigger payoff, namely, 9.

6.3 Subgame perfect equilibria and backward induction

When we find strategies in finite extensive-form games by backward induction, we are finding subgame perfect Nash equilibria. In fact, when we use backward induction, we are essentially considering every subgame in the entire game.

Strategies in a subgame perfect Nash equilibrium make sense no matter where in the game tree you use them. In contrast, at a Nash equilibrium that is not subgame perfect, at least one of the players is using a strategy that at some node tells her to make a move that would not be in her interest to make. For example, at a Nash equilibrium where a player is using a strategy that includes a threat, if the relevant node were reached, the strategy would tell the player to make a move that would hurt her. The success of such a strategy depends on this node not being reached!

There are some finite games in extensive form for which backward induction does not work. Recall the game in Figure 6.2, which we discussed in Section 1.4.

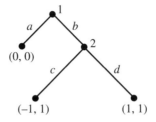

Figure 6.2. Failure of backward induction.

The problem with this game is that at the node where Player 2 chooses, both available moves give her a payoff of 1. Hence Player 1 does not know which move Player 2 will choose if that node is reached. However, Player 1 certainly wants to know which move Player 2 will choose before she decides between a and b!

In this game Player 1's strategy set is $S_1 = \{a, b\}$, and Player 2's strategy set is $S_2 = \{c, d\}$. In normal form, the game is just

		Player 2	
		c	d
Player 1	a	$(0, 0)$	$(0, 0)$
	b	$(-1, 1)$	$(1, 1)$

This game has two Nash equilibria, (a, c) and (b, d). Both are subgame perfect.

There is a way to find all subgame perfect Nash equilibria in any finite game in extensive form with perfect information by a variant of backward induction. Do backward induction as usual. If at any point a player has *several* best choices, record each of them as a possible choice at that point, and *separately* continue the backward induction using each of them. Ultimately you will find all subgame perfect Nash equilibria.

For example, in the game we are presently considering, we begin the backward induction at the node where Player 2 is to choose, since it is the only node all of whose successors are terminal. Player 2 has *two* best choices, c and d. Continuing the backward induction using c, we find that Player 1 chooses a. Continuing the backward induction using d, we find that Player 1 chooses b. Thus the two strategy profiles we find are (a, c) and (b, d). Both are subgame perfect Nash equilibria.

For a finite game in extensive form with complete information, this more general backward induction procedure never fails. Therefore *every finite game in extensive form with complete information has at least one subgame perfect Nash equilibrium.*

For a two-player zero-sum finite game in extensive form with complete information we know more (recall Section 5.5): any Nash equilibrium, and hence any subgame perfect Nash equilibrium, yields the same payoffs to the two players; and Player 1's strategy from one subgame perfect Nash equilibrium, played against Player 2's strategy from another subgame perfect Nash equilibrium, also yields those payoffs. Such strategies are "best" for the two players. This applies, for example, to chess: because of the rule that a game is a draw if a position is repeated three times, chess is a finite game. The game tree has about 10^{123} nodes. Since the tree is so large, best strategies for white (Player 1) and black (Player 2) are not known. In particular, it is not known whether the best strategies yield a win for white, a win for black, or a draw.

The notion of subgame perfect Nash equilibrium is especially valuable for infinite horizon games, for which backward induction cannot be used directly. These games are liable to have many Nash equilibria. Looking for the ones that are subgame perfect is a way of zeroing in on the (perhaps) most plausible ones. Sections 6.4–6.8 treat examples of infinite horizon games.

6.4 Duels and Truels

6.4.1 Duels. Two politicians, Newt and George, are competing for their party's presidential nomination. They take turns attacking each other. George is more skilled and better financed. Each time he attacks Newt, he has an 80% chance of knocking Newt out of the race. In contrast, each time Newt attacks George, he has a 50% chance of knocking George out of the race. The attacks continue until one candidate is driven from the race. If Newt attacks first, what is the probability that Newt is the survivor?

This problem is an example of a *duel*. It has several characteristics:

(1) The players take turns attacking.
(2) The duel could theoretically go on forever, if every attack fails to drive the opponent from the race.

In other duels, the players might attack simultaneously, so that a round could result in both players being driven from the race; this cannot happen when the players take turns. There are also duels that are not allowed to go on forever; it could be known in advance that the duel will end after a certain number of rounds if both candidates survive that long.

Duels are not games. The players have no need to choose strategies. However, we treat this duel much like a game. We will represent it by a tree diagram (see Figure 6.3). The terminal vertices are those where a player has just been eliminated. At the terminal vertices, we assign a payoff of 1 to the surviving candidate and 0 to the other candidate. We want to determine each candidate's expected payoff, which is just the probability the he is the ultimate survivor.

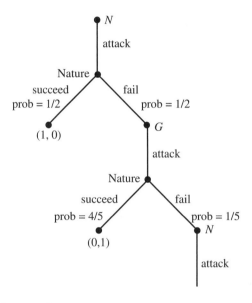

Figure 6.3. Tree diagram for the Duel. Player 1 is Newt (N), Player 2 is George (G).

Let π_1 denote Newt's expected payoff. If neither candidate is eliminated in the first two rounds, it will again be Newt's turn to attack. Since the situation will then be the same as it was at the start, Newt's expected payoff at that point will again be π_1. This leads to the formula

$$\pi_1 = \tfrac{1}{2} \times 1 + \tfrac{1}{2} \times \left(\tfrac{4}{5} \times 0 + \tfrac{1}{5} \times \pi_1\right).$$

Solving this equation for π_1 yields $\pi_1 = \tfrac{5}{9}$.

Thus Newt is the survivor $\tfrac{5}{9}$ of the time. Half the time he eliminates George with his first attack, and even if he fails, there is a chance he could eliminate George later.

6.4.2 Truels. A *truel* is like a duel except there are three contestants. Each time a contestant attacks, if he still has two surviving opponents, he must decide which opponent to attack. There are several types of truels:

(1) The contestants can take turns attacking or can attack simultaneously.

(2) The truel can be allowed to continue forever, or it can be required to end after a certain number of rounds.

(3) When it is a contestant's turn to attack and both opponents remain, he may be required to attack an opponent, or he may be allowed to choose not to attack at all.

A truel in which the contestants take turns ends when there is only one surviving contestant. A truel in which the contestants attack simultaneously also ends if all three contestants are eliminated.

Unlike a duel, a truel is a game, since the players have several available strategies to choose from. As with a duel, when two contestants have been eliminated, we assign a payoff of 1 to the surviving contestant and 0 to the others. Payoffs also must be assigned to the case in which the game continues forever. The payoffs that make sense in this case depend on the context.

6.4.3 Analysis of a truel. Three politicians, Newt, George, and Ron, are competing for their party's presidential nomination. They take turns attacking each other. When it is a candidate's turn to attack, if both opponents remain in the race, he chooses one to attack, or he can choose to attack neither. If only one opponent remains in the race, the candidate attacks him. When Newt attacks a candidate, he has a 50% chance of knocking him out of the race. When George attacks a candidate, he has an 80% chance of knocking him out of the race. Ron, however, is a legendary supercandidate. When Ron attacks someone, Ron is sure to knock him out of the race. The attacks continue until only one candidate is left. If Newt attacks first, George second, and Ron third, what are each candidate's chances of winning?

We denote the initial state of the game by NGR. This means that it is Newt's turn; he will be followed by George, then Ron (if they are still in the race). Because of the candidates' order, the other possible states of the game with three players left are GRN and RNG. There are six possible two-candidate states (NG, NR, GN, GR, RN, and RG), and of course three possible one-candidate states, which correspond to terminal vertices in the game tree. However, in drawing game trees, we regard the two-person states as terminal, since for each two-person state, we can calculate the expected payoff of each player. We did this for the two-person state NG when we analyzed the duel between Newt and George. Here are the expected payoffs for all two-person states; Players 1, 2, and 3 are Newt, George, and Ron, respectively:

$$(\pi_1, \pi_2, \pi_3)(NG) = \left(\tfrac{5}{9}, \tfrac{4}{9}, 0\right), \qquad (\pi_1, \pi_2, \pi_3)(NR) = \left(\tfrac{1}{2}, 0, \tfrac{1}{2}\right),$$

$$(\pi_1, \pi_2, \pi_3)(GN) = \left(\tfrac{1}{9}, \tfrac{8}{9}, 0\right), \qquad (\pi_1, \pi_2, \pi_3)(GR) = \left(0, \tfrac{4}{5}, \tfrac{1}{5}\right),$$

$$(\pi_1, \pi_2, \pi_3)(RN) = (0, 0, 1), \qquad (\pi_1, \pi_2, \pi_3)(RG) = (0, 0, 1).$$

The payoffs for *GN* are calculated like those for *NG*. The payoffs for two-person states involving Ron are easier to compute, since if he gets a turn he will certainly eliminate his remaining opponent.

A player's strategy is a plan for what to do each time he encounters the one three-person state where it is his turn to attack. For example, Newt needs a plan for each time the state *NGR* ("his" state) is encountered. A priori his planned move could depend on all players' previous moves. However, *we shall assume that each player plans the same move for every time his state is encountered.* For example, if Newt plans to attack Ron on his initial turn, and if the state *NGR* is arrived at later in the game, Newt will attack Ron then too. It is reasonable to expect that there will be a subgame perfect Nash equilibrium with strategies of this type.

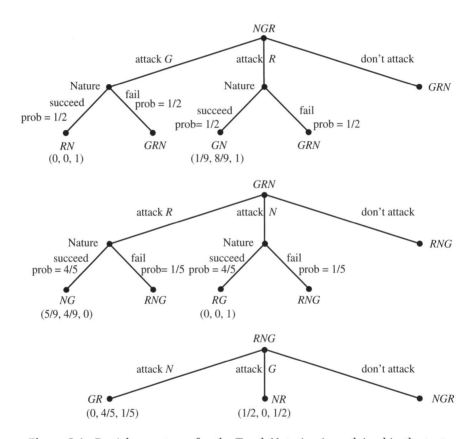

Figure 6.4. Partial game trees for the Truel. Notation is explained in the text.

Thus each player has only three strategies. For example, Newt can attack George (*G*), attack Ron(*R*), or not attack (∅) whenever the state *NGR* is encountered.

Figure 6.4 shows partial game trees starting at the three-person states. From these trees we see:

- In the situation NGR, Newt's strategy R dominates his strategy G.
- In the situation GRN, George's strategy R dominates his strategy N.
- In the situation RNG, Ron's strategy G dominates his strategy N.

In other words, if you are going to attack an opponent, you should attack the stronger one. The reason is obvious: if you leave only the weaker of your opponents, you have a better chance of surviving. We therefore eliminate one strategy for each candidate, leaving two for each.

Now we can regard this game as one in normal form with three candidates and two strategies each. The payoffs are shown in Table 6.1. The strategy profile $(\varnothing, \varnothing, \varnothing)$ results in all candidates surviving forever; we need to assign payoffs to this case. In this problem, the most reasonable payoffs to assign seem to be $(\frac{1}{3}, \frac{1}{3}, \frac{1}{3})$, since if no candidate is ever eliminated, each presumably has probability $\frac{1}{3}$ of winning the nomination.

Table 6.1. Payoff matrices for the Truel

Player 3 (Ron) attacks George

		Player 2 (George)	
		R	$\varnothing$
Player 1 (Newt)	R	$(\frac{59}{180}, \boxed{\frac{112}{180}}, \boxed{\frac{1}{20}})$	$(\frac{11}{36}, \frac{4}{9}, \boxed{\frac{1}{4}})$
	$\varnothing$	$(\boxed{\frac{49}{90}}, \boxed{\frac{16}{45}}, \boxed{\frac{1}{10}})$	$(\boxed{\frac{1}{2}}, 0, \boxed{\frac{1}{2}})$

Player 3 (Ron) does not attack

		Player 2 (George)	
		R	$\varnothing$
Player 1 (Newt)	R	$(\frac{25}{81}, \frac{56}{81}, 0)$	$(\frac{1}{9}, \boxed{\frac{8}{9}}, 0)$
	$\varnothing$	$(\boxed{\frac{5}{9}}, \boxed{\frac{4}{9}}, 0)$	$(\boxed{\frac{1}{3}}, \frac{1}{3}, \frac{1}{3})$

The table also shows best responses. There is one pure strategy Nash equilibrium. Newt, who starts the game, does not attack. The other two players each plan to attack their stronger opponent, who is not Newt. By thus lying low, Newt gets by far the best outcome of any player: with probability greater than half, he will be the lone survivor.

Some of the payoffs are easy to check. For example, if the strategy profile is $(R, \varnothing, G)$, Newt begins by attacking Ron. With probability $\frac{1}{2}$, Newt eliminates

Ron. The situation is now GN, which results in the payoffs $(\frac{1}{9}, \frac{8}{9}, 0)$. On the other hand, with probability $\frac{1}{2}$, Newt fails eliminate Ron. The situation is now GRN. Since George use the strategy $\varnothing$, he does not attack, and the situation becomes RNG. Since Ron uses the strategy G, he eliminates George, and the situation becomes NR. The payoffs from NR are $(\frac{1}{2}, 0, \frac{1}{2})$. Therefore

$$(\pi_1, \pi_2, \pi_3)(R, \varnothing, G) = \tfrac{1}{2}(\tfrac{1}{9}, \tfrac{8}{9}, 0) + \tfrac{1}{2}(\tfrac{1}{2}, 0, \tfrac{1}{2}) = (\tfrac{11}{36}, \tfrac{4}{9}, \tfrac{1}{4}).$$

Other payoffs can be checked by solving an equation. For example, if the strategy profile is $(R, R, \varnothing)$, Newt agains begins by attacking Ron. With probability $\frac{1}{2}$, Newt eliminates Ron, leading to the payoffs $(\frac{1}{9}, \frac{8}{9}, 0)$ as before. With probability $\frac{1}{2}$, Newt fails eliminate Ron, leading to the situation GRN. Since George now uses the strategy R, he attacks Ron. With probability $\frac{4}{5}$ he eliminates Ron, leading to the situation NG. This leads to the payoffs $(\frac{5}{9}, \frac{4}{9}, 0)$. With probability $\frac{1}{5}$, George fails to eliminate Ron. The situation becomes RNG. Since Ron uses the strategy $\varnothing$, he does not attack, and the situation becomes NGR: we are back where we started. Therefore

$$(\pi_1, \pi_2, \pi_3)(R, R, \varnothing) = \tfrac{1}{2}(\tfrac{1}{9}, \tfrac{8}{9}, 0) + \tfrac{1}{2}(\tfrac{4}{5}(\tfrac{5}{9}, \tfrac{4}{9}, 0) + \tfrac{1}{5}(\pi_1, \pi_2, \pi_3)).$$

The solution of this equation is $(\pi_1, \pi_2, \pi_3) = (\frac{25}{81}, \frac{56}{81}, 0)$.

We remark that we have found a Nash equilibrium under the assumption that each player is only allowed to use one of the three strategies we have described. Further analysis would be required to show that the strategy profile we have found remains a Nash equilibrium when other strategies are allowed and to show that it is subgame perfect.

6.5   The Rubinstein bargaining model

One dollar is to be split between two players. Player 1 goes first and offers to keep a fraction x_1 of the available money (one dollar). Of course, $0 \leqslant x_1 \leqslant 1$, and Player 2 would get the fraction $1 - x_1$ of the available money. If Player 2 accepts this proposal, the game is over, and the payoffs are x_1 to Player 1 and $1 - x_1$ to Player 2.

If Player 2 rejects the proposal, the money shrinks to δ dollars, $0 < \delta < 1$, and it becomes Player 2's turn to make an offer.

Player 2 offers a fraction y_1 of the available money (now δ dollars) to Player 1. Of course, $0 \leqslant y_1 \leqslant 1$, and Player 2 would get the fraction $1 - y_1$ of the available money. If Player 1 accepts this proposal, the game is over, and the payoffs are $y_1\delta$ to Player 1 and $(1 - y_1)\delta$ to Player 2.

If Player 1 rejects the proposal, the money shrinks to δ^2, and it becomes Player 1's turn to make an offer.

Player 1 offers to keep a fraction x_2 of the available money (now δ^2 dollars) and give the fraction $1 - x_2$ to Player 2. ...Well, you probably get the idea. See Figure 6.5.

The payoff to each player is the money she gets. If no proposal is ever accepted, the payoff to each player is 0.

This game models a situation in which it is in everyone's interest to reach an agreement quickly. Think, for example, of labor negotiations during a strike: as the strike goes on, the workers lose pay, and the company loses production. However, you probably don't want to reach a quick agreement by offering everything to your opponent!

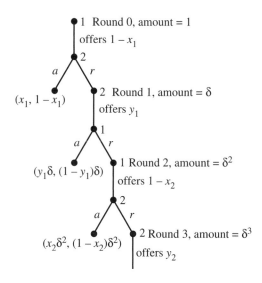

Figure 6.5. The Rubinstein bargaining model.

The numbering of the rounds of the game is shown in Figure 6.5.

A strategy for Player 1 consists of a plan for each round. For the even rounds, she must plan what offer to make. For the odd rounds, she must plan which offers to accept and which to reject. Of course, her plan can depend on what has happened up to that point.

Player 2's strategies are similar. For the even rounds, she must plan which offers to accept and which to reject. For the odd rounds, she must plan what offer to make.

Notice that at the start of any even round, Player 1 faces exactly the same situation that she faced at the start of the game, except that the available money is less. Similarly, at the start of any odd round, Player 2 faces exactly the same situation that she faced at the start of round 1, except that the available money is less.

We make two simplifying assumptions:

(1) Suppose a player has a choice between accepting an offer (thus terminating the game) and rejecting the offer (thus extending the game), and suppose the payoff the player expects from extending the game equals the offer she was just made. Then she will accept the offer, thus terminating the game.

(2) There is a subgame perfect Nash equilibrium with the following property: if it yields a payoff of x to Player 1, then in the subgame that starts at round 2, it yields a payoff of $x\delta^2$ to Player 1.

With these assumptions, in the subgame perfect equilibrium of assumption (2), if the game were to go to Round 2, the payoffs would be $(x\delta^2, (1-x)\delta^2)$. So the game tree gets pruned to the one shown in Figure 6.6.

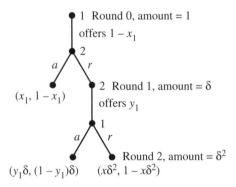

Figure 6.6. Pruned Rubinstein bargaining model.

Let us continue to investigate the subgame perfect equilibrium of assumption (2). Since it is subgame perfect, we should reason backward on the pruned game tree to find the players' remaining moves.

At round 1, Player 2 must make an offer. If she offers a fraction y_1^* of the available amount δ chosen so that $y_1^*\delta = x\delta^2$, Player 1 will be indifferent between accepting the offer and rejecting it. According to assumption 1, she will accept it. Player 2 will get to keep $\delta - y_1^*\delta$. If Player 2 offered more than y_1^*, Player 1 would accept, but Player 2 would get less than $\delta - y_1^*\delta$. If Player 2 offered less than y_1^*, Player 1 would not accept, and Player 2 would end up with $\delta^2 - x\delta^2$. This is less than $\delta - y_1^*\delta$:

$$\delta^2 - x\delta^2 = \delta^2 - y_1^*\delta < \delta - y_1^*\delta.$$

Thus Player 2's best move is to offer the fraction y_1^* of the available amount δ chosen so that $y_1^*\delta = x\delta^2$, that is, $y_1^* = x\delta$. Since this is her best move, it is the move she makes at round 1 in our subgame perfect equilibrium. Player 1 accepts the offer.

At round 0, Player 1 must make an offer. If she offers a fraction $1 - x_1^*$ of the available amount 1 chosen so that $1 - x_1^* = (1 - y_1^*)\delta$, Player 2 will be indifferent between accepting the offer and rejecting it. According to assumption (1), she will accept it. Player one gets to keep x_1^*. Reasoning as before, we see that if Player 1 offers more or less than $1 - x_1^*$, she ends up with less than x_1^*. Thus Player 1's best move is to offer the fraction $1 - x_1^*$, Player 2 accepts, and Player 1's payoff is x_1^*.

From assumption (2), we conclude that $x_1^* = x$.

From the equations $1 - x = 1 - x_1^* = (1 - y_1^*)\delta$ and $y_1^* = x\delta$, we obtain

$$1 - x = (1 - y_1^*)\delta = (1 - x\delta)\delta = \delta - x\delta^2,$$

so $1 - \delta = x - x\delta^2$, and so

$$x = \frac{1 - \delta}{1 - \delta^2} = \frac{1}{1 + \delta}. \tag{6.1}$$

This is Player1's payoff. Player 2's payoff is

$$1 - x = \frac{\delta}{1 + \delta}. \tag{6.2}$$

Player 1's payoff is higher. For δ close to 1 (i.e., when the money they are bargaining about does not shrink very fast), both payoffs are close to $\frac{1}{2}$.

6.6 Discount factor and repeated games

In Rubinstein's bargaining model, payoffs in the future are lower, because the amount of money to be divided decreases over time. However, there are good reasons to regard payoffs in the future as less valuable than payoffs today, even if they are not lower. One reason is that the future is always uncertain. A proverb says that a bird in the hand is worth two in the bush. Certainly a bird in the hand is worth more than one in the bush! A second reason is that money earns interest. If you can earn 4% annual interest, for example, a payoff of $100 today is surely worth more than a payoff of $100 in a year. If you had $100 today, you could invest it and have $104 in a year. That is why, if you win a big prize in a state lottery, you are offered either a certain amount now, or a larger amount over a period of years. A third reason is psychological. For most of us, the present is simply more important than the future. People differ, of course, in the relative importance they assign to the future.

In this section, we introduce the notion of discount factor, which formalizes the idea that payoffs in the future as less valuable than payoffs today. We then use the notion of discount factor to define repeated games, which will occupy us in the next two sections.

6.6.1 Discount factor. Suppose your boss proposes to you a salary of s dollars this year and t dollars next year. The utility to you of s dollars in the year you earn it is $u(s)$; recall the notion of a utility function introduced in Subsection 4.1.1. The total utility to you today of your boss's offer is $U(s,t) = u(s) + \delta u(t)$, where δ is a *discount factor* that is strictly between 0 and 1. The value of δ can reflect the uncertainty of the future, the interest rate you could earn, or the importance you attach to the future relative to the present.

Which would you prefer, a salary of s this year and s next year, or a salary of $s - a$ this year and $s + a$ next year? Assume $0 < a < s$, and, as we discussed in Subsection 4.1.1, assume $u' > 0$ (more is better) and $u'' < 0$ (the rate of increase of your utility decreases as your salary increases). Then

$$U(s,s) - U(s-a, s+a) = u(s) + \delta u(s) - (u(s-a) + \delta u(s+a))$$
$$= u(s) - u(s-a) - \delta(u(s+a) - u(s))$$
$$= \int_{s-a}^{s} u'(t)\, dt - \delta \int_{s}^{s+a} u'(t)\, dt > 0.$$

Hence you prefer s each year.

Do you see why the last line is positive? Part of the reason is that $u'(s)$ decreases as s increases, so $\int_{s-a}^{s} u'(t)\, dt > \int_{s}^{s+a} u'(t)\, dt$.

6.6.2 Definition of a repeated game. Let G be a game in normal form with players $1, \ldots, n$, strategy sets $S_1, \ldots, S_n$, and payoff functions $\pi_i(s_1, \ldots, s_n)$.

We define a *repeated game R* with *stage game G* and *discount factor δ*, $0 < \delta < 1$, as follows. The stage game G is played at times $k = 0, 1, 2, \ldots$, *with only pure strategies allowed*. A strategy for Player i is just a way of choosing which of his pure strategies s_i to use at each time k. His choice of which strategy to use at time k can depend on all the strategies used by all the players at times before k.

There will be a payoff to Player i from the stage game at each time $k = 0, 1, 2, \ldots$. His payoff in the repeated game R is just

his payoff at time $0 + \delta \times$ his payoff at time $1 + \delta^2 \times$ his payoff at time $2 + \cdots$.

In other words, payoffs at time 1 are discounted by a factor δ, payoffs at time 2 by the factor δ^2, and so forth.

A subgame of R is defined by taking the repeated game that starts at some time k. You should think of a subgame as including the memory of the strategies used by all players in the stage games at earlier times.

6.7   The Wine Merchant and the Connoisseur

A Wine Merchant sells good wine at $5 a bottle. The cost of this wine to the Wine Merchant is $4, so he makes $1 profit on each bottle he sells. Instead of doing this, the Wine Merchant could try to sell terrible wine at $4 a bottle. He can acquire terrible wine for essentially nothing.

Bernard is a regular customer and a wine connoisseur. He values the good wine at $6 a bottle, so when he buys it for $5, he feels he is ahead by $1. If he ever tasted the terrible wine, he would value it at 0.

Bernard can either pay for his wine or steal it. If he steals the good wine, the Wine Merchant will complain to everyone, which will result in a loss of reputation to Bernard worth $2. If Bernard steals the terrible wine, the Wine Merchant will not bother to complain.

We think of this situation as a two-player game in which the Wine Merchant has two strategies, sell good wine or sell bad wine. Bernard also has two strategies, pay for his wine or steal it.

Assuming Bernard wants one bottle of wine, we have the following game in normal form:

		Bernard	
		pay	steal
Wine Merchant	**sell good**	$(1, 1)$	$(-4, 4)$
	sell bad	$(4, -4)$	$(0, 0)$

The payoffs are explained as follows:

- $(1, 1)$: The Wine Merchant makes $1 profit when he sells good wine, Bernard gets a wine worth $6 for $5.
- $(-4, 4)$: The Wine Merchant loses a bottle of wine that cost him $4, Bernard gets a wine worth $6 at the cost of a $2 blow to his reputation.
- $(4, -4)$: The Wine Merchant makes $4 profit when he sells bad wine, Bernard has paid $4 for wine that is worth nothing to him.
- $(0, 0)$: The Wine Merchant loses a bottle of wine that cost him nothing, Bernard gets a wine worth nothing.

The Wine Merchant has a strictly dominant strategy: sell bad wine. Bernard also has a strictly dominant strategy: steal the wine. If both players use their nasty dominant strategies, each gets a payoff of 0. In contrast, if both players use their nice, dominated strategies (The Wine Merchant sells good wine, Bernard pays for it), both get a payoff of 1. This game is a prisoner's dilemma.

To make this game easier to discuss, let's call each player's nice strategy c for "cooperate," and let's call each player's nasty strategy d for "defect." Now we have the following payoff matrix:

		Bernard	
		c	d
Wine merchant	c	$(1, 1)$	$(-4, 4)$
	d	$(4, -4)$	$(0, 0)$

Both players have the same strategy set, namely, $\{c, d\}$.

We will take this game to be the stage game in a repeated game R. Recall that a player's strategy in a repeated game is a way of choosing which of his strategies to use at each time k. The choice can depend on what all players have done at times before k.

We consider the *trigger strategy* in this repeated game, which we denote σ and which is defined as follows. Start by using c. Continue to use c as long as your opponent uses c. If your opponent ever uses d, use d at your next turn, and continue to use d forever.

In other words, the Wine Merchant starts by selling good wine, but if his customer steals it, he decides to minimize his losses by selling terrible wine in the future. Bernard starts by buying from the Wine Merchant, but if the Wine Merchant cheats him, he says to himself, "I'm not going to pay that cheater any more!"

Theorem 6.1. *If $\delta \geqslant \frac{3}{4}$, then (σ, σ) is a Nash equilibrium.*

Proof. If both players use σ, then both cooperate in every round. Therefore both receive a payoff of 1 in every round. Taking into account the discount factor δ, we have

$$\pi_1(\sigma, \sigma) = \pi_2(\sigma, \sigma) = 1 + \delta + \delta^2 + \cdots = \frac{1}{1 - \delta}.$$

Here we have used the formula for the sum of an infinite geometric series:

$$r + r\delta + r\delta^2 + \cdots = \frac{r}{1 - \delta} \quad \text{provided } |\delta| < 1.$$

Suppose Bernard switches to a different strategy σ'. (Because of the symmetry of the game, the argument would be exactly the same for the Wine Merchant.)

Case 1. Bernard still ends up cooperating in every round. Then the Wine Merchant, who is still using σ, will also cooperate in every round. The payoffs are unchanged.

Case 2. Bernard first defects in round k. Then the Wine Merchant will cooperate through round k, will defect in round $k+1$, and will defect in every round after that. Does using σ' improve Bernard's payoff?

The payoffs from the strategy profiles (σ, σ) and (σ, σ') are the same through round $k - 1$, so let's just compare their payoffs to Bernard from round k on.

With (σ, σ), Bernard's payoff from round k on is

$$\delta^k + \delta^{k+1} + \cdots = \frac{\delta^k}{1 - \delta}.$$

With (σ, σ'), Bernard's payoff in round k is 4: the payoff from stealing good wine. From round $k + 1$ on, unfortunately, the Wine Merchant will defect (sell bad wine) in every round. Bernard's best response to this is to steal it, giving him a payoff of 0. Therefore, taking into account the discount factor, Bernard's payoff from round k on is at most

$$4\delta^k + 0(\delta^{k+1} + \delta^{k+2} + \cdots) = 4\delta^k.$$

From round k on, Bernard's payoff from (σ, σ) is greater than or equal to his payoff from (σ, σ') provided

$$\frac{\delta^k}{1 - \delta} \geqslant 4\delta^k \quad \text{or} \quad \delta \geqslant \frac{3}{4}.$$

□

Some people consider this way of analyzing repeated games to be game theory's most important contribution. It is said to show that cooperative behavior can establish itself in a society without the need for an external enforcer. In particular, it shows that cooperating in a prisoner's dilemma, as long as the other player does, can be rational when the game is to be repeated indefinitely. The lower bound for δ can be interpreted as meaning that that this conclusion holds when the players value the future highly enough.

Instead of using a discount factor, one can think of δ as the probability that the game will be repeated, given that it has just been played. The mathematics is exactly the same. The interpretation now is that if the probability of the game's being repeated is high enough, then it can be rational to use the cooperative strategy in a prisoner's dilemma.

Is (σ, σ) a subgame perfect Nash equilibrium? To answer this question, we must be sure we understand what σ tells us to do in every conceivable situation, including those in which σ has not been used correctly up to that point. For example, suppose your opponents defects, so you begin defecting, but in round j you make a mistake and accidentally cooperate. Should you go back to defecting in round $j + 1$, or should you check whether your opponent perhaps cooperated in round j, and base your decision on that? We have not defined σ carefully enough to answer this question.

Let's redefine the trigger strategy σ as follows. Start by using c. In round j, if both you and your opponent have used c in all previous rounds, use c. Otherwise, use d.

You should be able to convince yourself that with this precise definition, (σ, σ) is a subgame perfect Nash equilibrium. Remember that a subgame is defined by starting the game at some time, with the memory of all that has happened previously. For any such subgame, you should convince yourself that if your opponent is using the trigger strategy σ as we have just redefined it, you can do no better than to use σ yourself.

6.8 The Folk Theorem

Are there other Nash equilibria in the repeated game of Wine Merchant and Connoisseur? The so-called Folk Theorem of Repeated Games says that there are many. (In mathematics, a "folk theorem" is a theorem that lots of people seem to know, but no one knows who first proved it.)

6.8.1 Payoff vectors. For a game in normal form with n players, let $\sigma = (\sigma_1, \ldots, \sigma_n)$ be a mixed strategy profile. The *payoff vector* associated with σ is the n-tuple of numbers $(\pi_1(\sigma), \ldots, \pi_n(\sigma))$. An n-tuple of numbers $(v_1, \ldots, v_n)$ is a *possible payoff vector* for the game if it is the payoff vector associated with some strategy profile σ.

For example, consider a two-player game in normal form in which each player has two strategies:

		Player 2	
		t_1	t_2
Player 1	s_1	(a, b)	(c, d)
	s_2	(e, f)	(g, h)

If Player 1 uses the strategy $\sigma_1 = ps_1 + (1 - p)s_2$ with $0 \leqslant p \leqslant 1$, and Player 2 uses the strategy $\sigma_2 = qt_1 + (1 - q)t_2$ with $0 \leqslant q \leqslant 1$, then the payoff vector associated with the strategy profile (σ_1, σ_2) is

$$pq(a, b) + p(1 - q)(c, d) + (1 - p)q(e, f) + (1 - p)(1 - q)(g, h). \quad (6.3)$$

The set of all possible payoff vectors for the game can be thought of as a set of points in the plane. This set of points can be drawn as follows:

(1) Draw the line segment from (a, b) to (c, d). Think of it as parametrized by q, $0 \leqslant q \leqslant 1$, as follows: $r(q) = q(a, b) + (1 - q)(c, d)$.

(2) Draw the line segment from (e, f) to (g, h). Think of it as parametrized by q, $0 \leqslant q \leqslant 1$, as follows: $s(q) = q(e, f) + (1 - q)(g, h)$.

(3) For each q, draw the line segment from $r(q)$ to $s(q)$. This set of points is parameterized by p, $0 \leqslant p \leqslant 1$, as follows:

$$pr(q) + (1 - p)s(q) = p(q(a, b) + (1 - q)(c, d))$$
$$+ (1 - p)(q(e, f) + (1 - q)(g, h)).$$

Compare (6.3).

(4) The union of all the lines you have drawn is the set of all possible payoff vectors for the game.

Figure 6.7 shows an example.

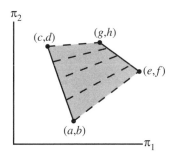

Figure 6.7. A set of possible payoff vectors. Typical line segments from $r(q)$ to $s(q)$ are dashed, and the set of all possible payoff vectors is gray.

6.8.2 Minimax payoffs. For a two-player game in normal form, Player 1's *minimax payoff* is the lowest payoff she ever gets when she makes her best response to one of Player 2's *pure* strategies. Player 1's minimax payoff is the lowest that Player 2 can force Player 1 to accept when using pure strategies. Player 2's minimax payoff is defined analogously

To state the definition of minimax payoff symbolically, let s_i denote one of Player 1's pure strategies, and let t_j denote one of Player 2's pure strategies. Then Player 1's minimax payoff is $\min_{t_j} \max_{s_i} \pi_1(s_i, t_j)$. Player 2's minimax payoff is $\min_{s_i} \max_{t_j} \pi_2(s_i, t_j)$.

For example, in the game of Wine Merchant and Connoisseur (Section 6.7), if Bernard uses c, the Wine Merchant's best response is d, which gives him a payoff of 4. If Bernard uses d, the Wine Merchant's best response is again d, which gives him a payoff of 0. The minimum of 4 and 0 is 0, so the Wine Merchant's minimax payoff is 0. Bernard can force the Wine Merchant to accept a payoff of 0 by using d, but he cannot force the Wine Merchant to accept any lower payoff. This fact was used in Bernard's trigger strategy to encourage the Wine Merchant to use c. If the Wine Merchant used d instead, Bernard would retaliate by using d, which would force the Wine Merchant to accept a payoff of 0. Since 0 is the Wine Merchant's minimax payoff, this is the worst punishment that Bernard can inflict.

6.8.3 The Folk Theorem.
One version of the Folk Theorem says the following.

Theorem 6.2. *The Folk Theorem. Let G be a two-player game in normal form, let m_1 be Player 1's minimax payoff in G, and let m_2 be Player 2's minimax payoff in G. Let R be the repeated game with stage game G and discount factor δ. Then:*

(1) At a Nash equilibrium of R, Player 1's payoff in every round is at least m_1, and Player 2's payoff in every round is at least m_2.

(2) Let (v_1, v_2) be a possible payoff vector for G with $v_1 > m_1$ and $v_2 > m_2$. Then there is a Nash equilibrium of R in which the average payoff vector is (v_1, v_2), provided the discount factor δ is large enough.

The average payoff vector is defined by taking the average payoff vector for the first k rounds, and then taking the limit as $k \to \infty$.

To show conclusion (1), we will just consider Player 1's payoff. Suppose we have arrived at round k of the repeated game. Player 2's strategy requires him to look at the history so far and choose a certain pure strategy t_j to use in round k. But then Player 1 can use the following strategy for round k: look at the history so far and choose the best response to t_j. This will give Player 1 a payoff of at least m_1 in round j.

We will not show (2) in general; we will only consider the case in which Players 1 and 2 each have two pure strategies. Let σ_1 and σ_2 be the strategies in G such that $\pi_1(\sigma_1, \sigma_2) = v_1$ and $\pi_2(\sigma_1, \sigma_2) = v_2$. Of course there are numbers p and q between 0 and 1 such that $\sigma_1 = ps_1 + (1 - p)s_2$ and $\sigma_2 = qt_1 + (1 - q)t_2$.

We only consider the case in which p and q are rational. We write these fractions as $p = \frac{k}{m}$ and $q = \frac{\ell}{n}$ with k, ℓ, m, and n integers.

We first describe a strategy $\tilde{\sigma}_2$ for Player 2 in the repeated game:

- First $q \cdot mn$ rounds: use t_1. Note that $q \cdot mn = \frac{\ell}{n} \cdot mn = \ell m$, which is an integer, as it should be.
- Next $(1 - q) \cdot mn = mn - \ell m$ rounds: use t_2.

We have described Player 2's strategy for the first mn rounds. He does the same thing for the next mn rounds, and in fact forever.

We now describe a strategy $\tilde{\sigma}_1$ for Player 1 in the repeated game:

- First $p \cdot q \cdot mn$ rounds: use s_1. Note that $p \cdot q \cdot mn = \frac{k}{m} \cdot \frac{\ell}{n} \cdot mn = k\ell$, which is an integer, as it should be.
- Next $(1 - p) \cdot q \cdot mn = \ell m - k\ell$ rounds: use s_2.
- Next $p \cdot (1 - q) \cdot mn$ rounds: use s_1. Note that $p \cdot (1 - q) \cdot mn = kn - k\ell$, which is an integer, as it should be.
- Next $(1 - p) \cdot (1 - q) \cdot mn = mn - kn - \ell m + k\ell$ rounds: use s_2.

We have described Player 1's strategy for the first mn rounds. He does the same thing for the next mn rounds, and in fact forever.

Player 1's total payoff for the first mn rounds is

$$pqmn\pi_1(s_1, t_1) + (1 - p)qmn\pi_1(s_2, t_1) + p(1 - q)mn\pi_1(s_1, t_2)$$
$$+ (1 - p)(1 - q)mn\pi_1(s_2, t_2).$$

To get Player 1's average payoff for the first mn rounds, divide by mn:

$$pq\pi_1(s_1, t_1) + (1 - p)q\pi_1(s_2, t_1)$$
$$+ p(1 - q)\pi_1(s_1, t_2) + (1 - p)(1 - q)\pi_1(s_2, t_2)$$
$$= \pi_1(ps_1 + (1 - p)s_2, qt_1 + (1 - q)t_2)$$
$$= \pi_1(\sigma_1, \sigma_2)$$
$$= v_1.$$

Similarly, Player 2's average payoff for the first mn rounds is $\pi_2(\sigma_1, \sigma_2) = v_2$, so the average payoff vector over the first mn rounds is (v_1, v_2). This plus the repetitive character of $\tilde{\sigma}_1$ and $\tilde{\sigma}_2$ imply that the average payoff vector for the repeated game from the strategy profile $(\tilde{\sigma}_1, \tilde{\sigma}_2)$ is also (v_1, v_2).

Now we define trigger strategies for Players 1 and 2 in the repeated game. Player 1: Start by using $\tilde{\sigma}_1$. If Player 2 ever deviates from his strategy $\tilde{\sigma}_2$, use the strategy s_i to which Player 2's best response gives him payoff $m_2 < v_2$, and continue to use it forever. Player 2: Start by using $\tilde{\sigma}_2$. If Player 1 ever deviates from his strategy $\tilde{\sigma}_1$, use the strategy t_j to which Player 1's best response gives him payoff $m_1 < v_1$, and continue to use it forever. The profile of these strategies is the desired Nash equilibrium of the repeated game for sufficiently large δ. (Without a sufficiently large discount factor, Player 1, for example, could at some stage use his best response to Player 2's action at that stage, instead of using the action prescribed by his strategy $\tilde{\sigma}_1$. This would give him a better payoff at that stage; with a small enough discount factor, he would also have a better payoff in the repeated game.)

Is the profile of these trigger strategies subgame perfect? Only if s_i and t_j are best responses to each other. If they are, then if we get to the point where both players are using these strategies to punish the other, neither will be able to improve himself by changing on his own. This is what happens in prisoner's dilemma games, such as Wine Merchant and Connoisseur.

6.8.4 Wine Merchant and Connoisseur. For the game of Wine Merchant and Connoisseur, Figure 6.8 shows the possible payoff vectors (v_1, v_2) with $v_1 > 0$ and $v_2 > 0$ (since 0 is the minimax payoff for both players). According to the Folk Theorem, for any (v_1, v_2) in this set, there is a Nash equilibrium of the repeated game that gives payoffs (v_1, v_2) in every round, provided the discount factor δ is large enough.

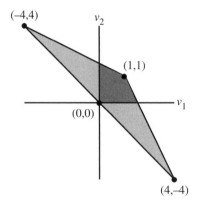

Figure 6.8. Possible payoff vectors for the game of Wine Merchant and Connoisseur (light gray). Those with $v_1 > 0$ and $v_2 > 0$ are in dark gray.

6.9 Maximum value of a function

Sections 6.10 and 6.11 discuss uses of backward induction by two Nobel prize-winning economists. To prepare for these examples, in this section we review some facts about the maximum value of a function.

Suppose f is a continuous function on an interval $a \leqslant x \leqslant b$. From calculus we know:

(1) f attains a maximum value somewhere on the interval.
(2) The maximum value of f occurs at a point where $f' = 0$, or at a point where f' does not exist, or at an endpoint of the interval.
(3) If $f'(a) > 0$, the maximum does not occur at a.
(4) If $f'(b) < 0$, the maximum does not occur at b.

Suppose that $f'' < 0$ everywhere in the interval $a \leqslant x \leqslant b$. Then we know a few additional things:

(1) f attains its maximum value at *unique* point c in $[a, b]$.
(2) Suppose $f'(x_0) > 0$ at some point $x_0 < b$. Then $x_0 < c$. See Figure 6.9.
(3) Suppose $f'(x_1) < 0$ at some point $x_1 > a$. Then $c < x_1$.

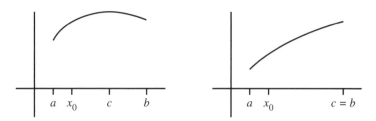

Figure 6.9. Two functions on $[a, b]$ with negative second derivative everywhere and positive first derivative at a point $x_0 < b$. Such functions always attain their maximum at a point c to the right of x_0.

6.10 The Samaritan's Dilemma

There is someone you want to help should she need it. However, you are worried that the very fact that you are willing to help may lead her to do less for herself than she otherwise would. This is the Samaritan's Dilemma.

The Samaritan's Dilemma is an example of *moral hazard*. Moral hazard is "the prospect that a party insulated from risk may behave differently from the way it would behave if it were fully exposed to the risk." There is a Wikipedia article on moral hazard: http://en.wikipedia.org/wiki/Moral_hazard.

Here is an example of the Samaritan's Dilemma analyzed by James Buchanan (Nobel Prize in Economics, 1986; Wikipedia article http://en.wikipedia.org/wiki/James_M._Buchanan).

A young woman plans to go to college next year. This year she is working and saving for college. If she needs additional help, her father will give her some of the money he earns this year.

Notation and assumptions regarding income and savings are as follows:

(1) Father's income this year is $z > 0$, which is known. Of this he will give $0 \leqslant t \leqslant z$ to his daughter next year.
(2) Daughter's income this year is $y > 0$, which is also known. Of this she saves $0 \leqslant s \leqslant y$ to spend on college next year.
(3) Daughter chooses the amount s of her income to save for college. Father then observes s and chooses the amount t to give to his daughter.

The important point is (3): after Daughter is done saving, Father will choose an amount to give to her. Thus Daughter, who goes first in this game, can use backward induction to figure out how much to save. In other words, she can take into account that different savings levels will result in different levels of support from Father.

The Utility functions (see Subsection 4.1.1) are

(1) Daughter's utility function $\pi_1(s,t)$, which is her payoff in this game, is the sum of

  (a) her first-year utility v_1, a function of the amount she has to spend in the first year, which is $y - s$; and

  (b) her second-year utility v_2, a function of the amount she has to spend in the second year, which is $s + t$. Second-year utility is multiplied by a discount factor $\delta > 0$.

  Thus we have

$$\pi_1(s,t) = v_1(y - s) + \delta v_2(s + t).$$

(2) Father's utility function $\pi_2(s,t)$, which is his payoff in this game, is the sum of

  (a) his personal utility u, a function of the amount he has to spend in the first year, which is $z - t$; and

  (b) his daughter's utility π_1, multiplied by a "coefficient of altruism" $\alpha > 0$.

  Thus we have

$$\pi_2(s,t) = u(z - t) + \alpha \pi_1(s,t) = u(z - t) + \alpha(v_1(y - s) + \delta v_2(s + t)).$$

Notice that a component of Father's utility is Daughter's utility. The Samaritan's Dilemma arises when the welfare of someone else is important to us.

We assume (recall Subsection 4.1.1)

(A1) The functions v_1, v_2, and u have positive first derivatives and negative second derivatives.

Let's first gather some facts that we will use in our analysis.

(1) Formulas we will need for partial derivatives. They are calculated using the chain rule of first-semester calculus:

$$\frac{\partial \pi_1}{\partial s}(s,t) = -v_1'(y - s) + \delta v_2'(s + t),$$

$$\frac{\partial \pi_2}{\partial t}(s,t) = -u'(z - t) + \alpha \delta v_2'(s + t).$$

(2) Formulas we will need for second partial derivatives:

$$\frac{\partial^2 \pi_1}{\partial s^2}(s,t) = v_1''(y-s) + \delta v_2''(s+t),$$

$$\frac{\partial^2 \pi_2}{\partial s \partial t}(s,t) = \alpha \delta v_2''(s+t),$$

$$\frac{\partial^2 \pi_2}{\partial t^2}(s,t) = u''(z-t) + \alpha \delta v_2''(s+t).$$

All three of these are negative everywhere.

To find a savings level for Daughter using backward induction, we must first figure out, for each possible savings level s of Daughter, the contribution t that Father will make. To do this, we maximize Father's utility $\pi_2(s,t)$ with s fixed and $0 \leqslant t \leqslant z$ (recall that z is Father's income). Let's keep things from getting too complicated by arranging that for s fixed, $\pi_2(s,t)$ will attain its maximum at some t strictly between 0 and z. In other words, no matter how much Daughter saves, Father will give her some of his income but not all. This is guaranteed to happen if $(\partial \pi_2 / \partial t)(s,0) > 0$ and $(\partial \pi_2 / \partial t)(s,z) < 0$. The first condition prevents Father from giving Daughter nothing. The second prevents him from giving Daughter everything.

For $0 \leqslant s \leqslant y$, we have

$$\frac{\partial \pi_2}{\partial t}(s,0) = -u'(z) + \alpha \delta v_2'(s) \geqslant -u'(z) + \alpha \delta v_2'(y)$$

and

$$\frac{\partial \pi_2}{\partial t}(s,z) = -u'(0) + \alpha \delta v_2'(s+z) \leqslant -u'(0) + \alpha \delta v_2'(z).$$

We therefore make two more assumptions:

(A2) $\alpha \delta v_2'(y) > u'(z)$. This assumption is reasonable. We expect Daughter's income y to be much less than Father's income z. Since, as discussed in Subsection 4.1.1, each dollar of added income is less important when income is higher, we expect $v_2'(y)$ to be much greater than $u'(z)$. If the product $\alpha \delta$ is not too small (meaning that Father cares quite a bit about Daughter, and Daughter cares quite a bit about the future), we get our assumption.

(A3) $u'(0) > \alpha \delta v_2'(z)$. This assumption is reasonable because $u'(0)$ should be large and $v_2'(z)$ should be small.

With these assumptions, we have

$$\frac{\partial \pi_2}{\partial t}(s,0) > 0 \quad \text{and} \quad \frac{\partial \pi_2}{\partial t}(s,z) < 0 \quad \text{for all } 0 \leqslant s \leqslant y.$$

Since $\partial^2 \pi_2 / \partial t^2$ is always negative, from Section 6.9, there is a single value of t where $\pi_2(s,t)$, s fixed, attains its maximum value; moreover, $0 < t < z$, so

$(\partial \pi_2/\partial t)(s,t) = 0$ at this value of t. We denote this value of t by $t = b(s)$. This is Father's best-response strategy, the amount Father will give to Daughter if the amount Daughter saves is s.

Daughter now chooses her saving level $s = s^*$ to maximize the function $\pi_1(s, b(s))$, which we shall denote $V(s)$:

$$V(s) = \pi_1(s, b(s)) = v_1(y - s) + \delta v_2(s + b(s)).$$

Father then contributes $t^* = b(s^*)$.

Here is the punchline: suppose it turns out that $0 < s^* < y$; that is, Daughter saves some of her income but not all. (This is the usual case.) Then, had Father simply committed himself in advance to providing t^* in support to his daughter no matter how much she saved, Daughter would have chosen a savings rate $s^\sharp$ greater than s^*. *Both* Daughter and Father would have ended up with higher utility.

To see this, note that we have

$$\frac{\partial \pi_1}{\partial s}(s^*, t^*) = -v_1'(y - s^*) + \delta v_2'(s^* + t^*). \tag{6.4}$$

Suppose that we can show that this expression is positive. Then, since $(\partial^2 \pi_1/\partial s^2)(s, t^*)$ is always negative, we have that $\pi_1(s, t^*)$ is maximum at a value $s = s^\sharp$ greater than s^*. (See Section 6.9.)

We of course have $\pi_1(s^\sharp, t^*) > \pi_1(s^*, t^*)$, so Daughter's utility is higher. Since Daugher's utility is higher, we see from the formula for π_2 that $\pi_2(s^\sharp, t^*) > \pi_2(s^*, t^*)$, so Father's utility is also higher.

However, it is not obvious that (6.4) is positive. To see that it is, we proceed as follows.

Step 1. To maximize $V(s)$, we calculate

$$V'(s) = -v_1'(y - s) + \delta v_2'(s + b(s))(1 + b'(s)).$$

Step 2. If $V(s)$ is maximum at $s = s^*$ with $0 < s^* < y$, we must have $V'(s^*) = 0$, that is,

$$0 = -v_1'(y - s^*) + \delta v_2'(s^* + t^*)(1 + b'(s^*)). \tag{6.5}$$

Step 3. Subtracting (6.5) from (6.4), we obtain

$$\frac{\partial \pi_1}{\partial s}(s^*, t^*) = -\delta v_2'(s^* + t^*)b'(s^*). \tag{6.6}$$

Step 4. We expect that $b'(s) < 0$; this simply says that if Daughter saves more, Father will contribute less. To check this, note that

$$\frac{\partial \pi_2}{\partial t}(s, b(s)) = 0 \quad \text{for all } s.$$

Differentiating both sides of this equation with respect to s, we get

$$\frac{\partial^2 \pi_2}{\partial s \partial t}(s, b(s)) + \frac{\partial^2 \pi_2}{\partial t^2}(s, b(s))b'(s) = 0.$$

Since $\partial^2 \pi_2/\partial s \partial t$ and $\partial^2 \pi_2/\partial t^2$ are always negative, we must have $b'(s) < 0$.

Step 5. From (6.6), since v_2' is always positive and $b'(s)$ is always negative, we see that $(\partial \pi_1/\partial s)(s^*, t^*)$ is positive.

This problem has implications for government social policy. It suggests that social programs be made available to everyone rather than on an if-needed basis.

Let's look more closely at this conclusion.

When Father promises Daughter a certain fixed amount of help, one can imagine two possible effects: (i) now that she knows she will get this help, Daughter will save less; (ii) now that more saving will not result in less contribution from Father (remember, $b'(s) < 0$), Daughter will save more. All we have shown is that *if the promised contribution is t^**, it is actually (ii) that will occur. Too great a promised contribution might result in (i) instead.

In addition, our conclusion required that the coefficient of altruism α not be too small. That makes sense for a father and daughter. Whether it is correct for rich people (who do most of the paying for social programs) and poor people (who get most of the benefits) is less certain.

6.11 The Rotten Kid Theorem

A rotten son manages a family business. The amount of effort the son puts into the business affects both his income and his mother's. The son, being rotten, cares only about his own income, not his mother's. To make matters worse, Mother dearly loves her son. If the son's income is low, Mother will give part of her own income to her son so that he will not suffer. In this situation, can the son be expected to do what is best for the family?

We shall give the analysis of Gary Becker (Nobel Prize in Economics, 1992; Wikipedia article http://en.wikipedia.org/wiki/Gary_Becker).

We denote the son's annual income by y and the mother's by z. The amount of effort that the son devotes to the family business is denoted by a. His choice of a will affect both his income and his mother's, so we regard both y and z as functions of a: $y = y(a)$ and $z = z(a)$.

After Mother observes a, and hence observes her own income $z(a)$ and her son's income $y(a)$, she chooses an amount t, $0 \leqslant t \leqslant z(a)$, to give to her son.

The mother and son have personal utility functions u and v, respectively (see Subsection 4.1.1). Each is a function of the amount they have to spend.

The son chooses his effort a to maximize his own utility v, without regard for his mother's utility u. Mother, however, chooses the amount t to transfer to her son to maximize $u(z - t) + \alpha v(y + t)$, where α is her coefficient of altruism. Thus the payoff functions for this game are

$$\pi_1(a, t) = v(y(a) + t),$$
$$\pi_2(a, t) = u(z(a) - t) + \alpha v(y(a) + t).$$

Since the son chooses first, he can use backward induction to decide how much effort to put into the family business. In other words, he can take into account that even if he doesn't put in much effort, and so doesn't produce much income for either himself or his mother, his mother will help him out.

We make the following assumptions:

- As in Subsection 4.1.1, the functions u and v have positive first derivatives and negative second derivatives.
- The son's level of effort is chosen from an interval $I = [a_1, a_2]$.
- For all a in I, $\alpha v'(y(a)) > u'(z(a))$. This assumption expresses two ideas: (a) Mother dearly loves her son, so α is not small; and (b) no matter how little or how much the son works, Mother's income $z(a)$ is much larger than son's income $y(a)$. (Recall that the derivative of a utility function gets smaller as the income gets larger.) This makes sense if the income generated by the family business is small compared to Mother's overall income
- For all a in I, $u'(0) > \alpha v'(y(a) + z(a))$. This assumption is reasonable, because $u'(0)$ should be large and $v'(y(a) + z(a))$ should be small.
- Let $T(a) = y(a) + z(a)$ denote total family income. Then $T'(a) = 0$ at a unique point $a^\sharp$, $a_1 < a^\sharp < a_2$, and $T(a)$ attains its maximum value at this point. This assumption expresses the idea that if the son works too hard, he will do more harm than good. As they say in the software industry, if you stay at work too late, you're just adding bugs.

To find the son's level of effort using backward induction, we must first maximize $\pi_2(a, t)$ with a fixed and $0 \leqslant t \leqslant z(a)$. We calculate

$$\frac{\partial \pi_2}{\partial t}(a, t) = -u'(z(a) - t) + \alpha v'(y(a) + t),$$

$$\frac{\partial \pi_2}{\partial t}(a, 0) = -u'(z(a)) + \alpha v'(y(a)) > 0,$$

$$\frac{\partial \pi_2}{\partial t}(a, z(a)) = -u'(0) + \alpha v'(y(a) + z(a)) < 0,$$

$$\frac{\partial^2 \pi_2}{\partial t^2}(a, t) = u''(z(a) - t) + \alpha v''(y(a) + t) < 0.$$

Then there is a single value of t where $\pi_2(a, t)$, a fixed, attains its maximum; moreover, $0 < t < z(a)$, so $(\partial\pi_2/\partial t)(a, t) = 0$. (See Section 6.9.) We denote this value of t by $t = b(a)$. This is Mother's strategy, the amount Mother will give to her son if his level of effort in the family business is a.

The son now chooses his level of effort $a = a^*$ to maximize the function $\pi_1(a, b(a))$, which we shall denote $V(a)$:

$$V(a) = \pi_1(a, b(a)) = v(y(a) + b(a)).$$

Mother then contributes $t^* = b(a^*)$.

So what? Here is Becker's point.

Suppose $a_1 < a^* < a_2$ (the usual case). Then $V'(a^*) = 0$, that is,

$$v'(y(a^*) + b(a^*))(y'(a^*) + b'(a^*)) = 0.$$

Since v' is positive everywhere, we have

$$y'(a^*) + b'(a^*) = 0. \tag{6.7}$$

Now $(\partial\pi_2/\partial t)(a, t) = 0$ when $t = b(a)$, so $-u'(z(a) - b(a)) + \alpha v'(y(a) + b(a)) = 0$ for all a. Differentiating this equation with respect to a, we find that, for all a,

$$-u''(z(a) - b(a))(z'(a) - b'(a)) + \alpha v''(y(a) + b(a))(y'(a) + b'(a)) = 0.$$

In particular, for $a = a^*$,

$$\begin{aligned}
- u''(z(a^*) - b(a^*))&(z'(a^*) - b'(a^*)) \\
&+ \alpha v''(y(a^*) + b(a^*))(y'(a^*) + b'(a^*)) = 0.
\end{aligned}$$

This equation and (6.7) imply that

$$z'(a^*) - b'(a^*) = 0.$$

Adding this equation to (6.7), we obtain

$$y'(a^*) + z'(a^*) = 0.$$

Therefore $T'(a^*) = 0$. But then, by our last assumption, $a^* = a^\sharp$, the level of effort that maximizes total family income.

Thus, if the son had not been rotten, and instead had been trying to maximize total family income $y(a) + z(a)$, he would have chosen the same level of effort a^*.

6.12 Problems

6.12.1 Another Debate. Redo the problem analyzed in Subsection 6.4.3 with the assumption that Newt eliminates the candidate he attacks 20% of the time, George eliminates the candidate he attacks 30% of the time, and Ron always eliminates the candidate he attacks. Find a pure strategy Nash equilibrium, and give each candidate's probability of survival at the Nash equilibrium

6.12.2 A Truel with Simultaneous Attacks. In the problem analyzed in Subsection 6.4.3, suppose that the candidates attack simultaneously instead of taking turns. The game now has just eight states: NGR (all candidates still competing); NG, NR, and GN (two candidates left); N, G, and R (one candidate left); and O (no candidates left). The last four states correspond to terminal vertices. The payoffs when no candidate is left are $(0, 0, 0)$.

We assume that in each two-person state, each remaining candidate attacks the other. In the three-person state, each candidate has three strategies: attack one of the other candidates, or attack neither.

(1) Calculate the expected payoffs in each two-person state. After you have done this, you can regard the two-person states as terminal.
(2) Calculate the expected payoffs from the strategy profile $(\varnothing, R, G)$ (Newt lies low).
(3) Calculate the expected payoffs from the strategy profile (R, R, G) (the weaker candidates gang up on the strongest).

6.12.3 Huey, Dewey, and Louie Split a Dollar. Huey (Player 1), Dewey (Player 2), and Louie (Player 3) have a dollar to split.

Round 0: Huey goes first and offers to split the dollar into fractions a_1 for himself, b_1 for Dewey, and c_1 for Louie, with $a_1 + b_1 + c_1 = 1$. If Dewey and Louie both accept, the game is over. If at least one rejects the offer, the dollar shrinks to δ, $0 < \delta < 1$, and it is Dewey's turn to offer.

Round 1: Dewey (Player 2) offers to split the remaining money into fractions d_1 for Huey, e_1 for himself, and f_1 for Louie, with $d_1 + e_1 + f_1 = 1$. If Huey and Louie both accept, the game is over. If at least one rejects the offer, the remaining money shrinks to δ^2, and it is Louie's turn to offer.

Round 2: Louie (Player 3) offers to split the remaining money into fractions g_1 for Huey, h_1 for Dewey, and k_1 for himself, with $g_1 + h_1 + k_1 = 1$. If Huey and Dewey both accept, the game is over. If at least one rejects the offer, the remaining money shrinks to δ^3, and it is Huey's turn to offer.

Round 3: Huey (Player 1) offers to split the remaining money into fractions a_2 for himself, b_2 for Dewey, and c_2 for Louie, with $a_2 + b_2 + c_2 = 1$. If Dewey and Louie both accept, the game is over. If at least one rejects the offer, the remaining money shrinks to δ^4, and it is Dewey's turn to offer.

And so forth.

A schematic game tree is shown in Figure 6.10.

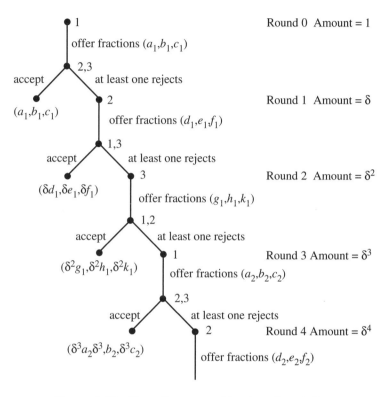

Figure 6.10. Huey, Dewey, and Louie split a dollar.

We make two simplifying assumptions:

- Suppose a player has a choice between accepting an offer and rejecting the offer, and suppose the offer equals the payoff the player expects from rejecting the offer. Then he will accept the offer.
- There is a subgame perfect Nash equilibrium with the following property: if it yields payoffs of x to Player 1, y to Player 2, and z to Player 3, then in the subgame that starts at round 3, it yields payoffs $\delta^3 x$ to Player 1, $\delta^3 y$ to Player 2, and $\delta^3 z$ to Player 3.

Because of the second assumption, we can prune the game tree to that shown in Figure 6.11.

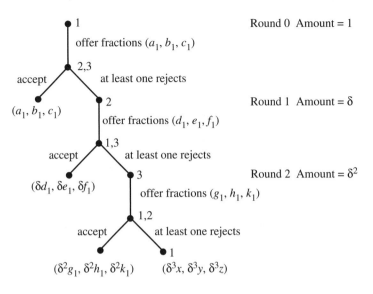

Figure 6.11. Pruned game tree.

(1) Explain why $g_1 = \delta x$ and $h_1 = \delta y$.

(2) Explain why $d_1 = \delta g_1$ and $f_1 = \delta k_1$.

(3) Explain why $b_1 = \delta e_1$ and $c_1 = \delta f_1$.

(4) Explain why $b_1 = y$ and $c_1 = z$.

(5) Parts (3) and (4) yield the two equations $y = \delta e_1$ and $z = \delta f_1$. Parts (1) and (2) yield four more equations. We also have the three equations

$$x + y + z = 1, \qquad d_1 + e_1 + f_1 = 1, \qquad g_1 + h_1 + k_1 = 1.$$

In total we have nine equations in the nine unknowns $x, y, z, d_1, e_1, f_1, g_1, h_1$, and k_1. Check that the following is a solution of these nine equations. (Actually, it's the only solution.)

$$x = e_1 = k_1 = \frac{1}{1 + \delta + \delta^2},$$
$$y = f_1 = g_1 = \frac{\delta}{1 + \delta + \delta^2},$$
$$z = d_1 = h_1 = \frac{\delta^2}{1 + \delta + \delta^2}.$$

6.12.4 Iran and Iraq. There are two oil-producing countries, Iran and Iraq. Both can operate at either of two production levels: low (2 million barrels per day) or high (4 million barrels per day). Depending on their decisions, total output will be 4, 6, or 8 million barrels per day. The price per barrel in the three cases will be $100, $60, or $40, respectively. Cost of production is $8 per barrel for Iran and $16 per barrel for Iraq. The following matrix shows profit per day in millions of dollars:

		Iraq low	Iraq high
Iran	low	$(184, 168)$	$(104, 176)$
	high	$(208, 88)$	$(128, 96)$

(1) Explain the numbers in the matrix.
(2) Find a discount factor δ_0 such that for $\delta \geqslant \delta_0$, it is a Nash equilibrium for both countries to use the following trigger strategy. Start by producing at the low level. If the other country produces at the high level even once, produce at the high level forever.

Remark: You will find a discount factor δ_1 such that for $\delta \geqslant \delta_1$, Iran cannot improve its payoff; and a second discount factor δ_2 such that for $\delta \geqslant \delta_2$, Iraq cannot improve its payoff. That means that for $\delta \geqslant \max(\delta_1, \delta_2)$, neither country can improve its payoff.

6.12.5 Should You Contribute? This problem is related to the Global Warming Game (Section 2.5). A group of ten students plays the following game. Each student is given one dollar. Each student then simultaneously puts a portion of her dollar into a pot. The game organizer counts the total amount in the pot, multiplies by five, and splits this amount equally among the ten students.

The ith student's strategy is a real number x_i, $0 \leqslant x_i \leqslant 1$, which represents the amount that student chooses to put in the pot. A strategy profile is therefore a 10-tuple $(x_1, x_2, \ldots, x_{10})$ with $0 \leqslant x_i \leqslant 1$ for each i.

(1) Find the ith player's payoff function $\pi_i(x_1, x_2, \ldots, x_{10})$. (The answer is

$$\pi_i(x_1, x_2, \ldots, x_{10}) = 1 - x_i + \tfrac{1}{2}(x_1 + x_2 + \cdots + x_{10});$$

you should explain this.)
(2) Show that each player has a strategy that strictly dominates all his other strategies: whatever the other players do, contribute nothing. (For example, consider Player 1. Given any choices $(x_2, \ldots, x_{10})$ by the other players, Player 1 maximizes her payoff by choosing $x_1 = 0$.) Therefore the only Nash equilibrium is $(0, 0, \ldots, 0)$, at which each player's payoff is 1.

(3) Suppose the game is repeated every day. Consider the following strategy σ_x, where $0 < x \leqslant 1$: "I will contribute x dollars on day 0. If every other player contributes at least x dollars on day k, I will contribute x dollars on day $k + 1$. If any player contributes less than x dollars on day k, I will contribute nothing on every subsequent day." Show that if $\delta \geqslant \frac{1}{9}$, then it is a Nash equilibrium for every player to use the strategy σ_x with the same x. (In other words, $(\sigma_x, \sigma_x, \ldots, \sigma_x)$ is a Nash equilibrium.)

6.12.6 Tit for Tat 3. We saw tit for tat in a game repeated twice in Problem 3.12.6. In the infinitely repeated game of Wine Merchant and Connoisseur, or in any prisoner's dilemma, tit for tat is the following strategy. Start by using c. Thereafter do whatever your opponent did in the previous round.

Suppose that in Wine Merchant and Connoisseur, both players use tit for tat. We will find the payoffs and check whether various alternative strategies by one of the players improve his payoff.

(1) "If both players use tit for tat, their payoffs are the same as when both players use the trigger strategy of Section 6.7." Explain.
(2) Suppose Player 2 uses tit for tat, but Player 1 uses the following variant: "Start by using d. Thereafter do whatever your opponent did in the previous round." Find δ_0 such that for $\delta \geqslant \delta_0$, Player 1 does not improve his payoff by using this variant.
(3) Suppose Player 2 uses tit for tat, but Player 1 uses d in every round. Find δ_0 such that for $\delta \geqslant \delta_0$, Player 1 does not improve his payoff by using this strategy.
(4) The previous two parts of this problem suggest that for δ sufficiently large, it is a Nash equilibrium of the repeated game for both players to use tit for tat. This is true, but we shall not complete the argument. Instead, show the following: it is *not* a subgame perfect Nash equilibrium of the repeated game for both players to use tit for tat. Suggestion: Suppose in period 0, Player 1 uses d and Player 2 uses c. Consider the subgame that begins with period 1, with the memory of what happened in period 0. What happens if both players use tit for tat in the subgame? What happens if Player 1 uses tit for tat in the subgame, but Player 2 "forgives" Player 1 by using c in round 1 and tit for tat thereafter? Find δ_0 such that for $\delta > \delta_0$, the second outcome gives Player 2 a better result than the first in the subgame.

Tit for tat has the drawback that, if one player defects, even by accident or for reasons beyond his control, the players become trapped in a cycle of repeated defections, as in part 2 of this problem. Israeli-Palestinian relations (Section 2.4) have been plagued by such cycles of repeated revenge-taking.

An alternative to tit for tat is "forgiving tit for tat," which is tit for tat except that, with some probability, a player responds to a defection by forgiving it and cooperating. Part 4 of this problem shows how forgiving tit for tat can be advantageous.

For more information on tit for tat, see the Wikipedia page http://en.wiki pedia.org/wiki/Tit_for_tat.

Chapter 7

Symmetries of games

In some games, players or strategies are interchangeable in some way. Nash discussed such games in the same paper [11] in which he proved his theorem on existence of Nash equlibria (Theorem 5.1). Nash showed that among the Nash equilibria of such games are some particularly simple ones. Finding such Nash equilibria can be much easier than finding all Nash equilibria.

In this chapter we mainly discuss games in which all players are interchangeable. In the last section we discuss some other situations.

7.1 Interchangeable players

We have seen several examples of games in normal form in which the players are interchangeable:

- Prisoner's Dilemma
- Stag Hunt
- Chicken
- Battle of the Sexes
- Water Pollution
- Tobacco Market
- Cournot Duopoly

For each of these games except Chicken and Battle of the Sexes, we found at least one pure strategy Nash equilibrium in which every player used the same strategy.

Recall from Subsection 3.2.3 the payoffs in Chicken:

		Teenager 2	
		straight	swerve
Teenager 1	straight	$(-2, -2)$	$(1, -1)$
	swerve	$(-1, 1)$	$(0, 0)$

There are two Nash equilibria: (straight, swerve) and (swerve, straight). In both of these Nash equilibria, of course, the players use different strategies.

Let's look for a mixed strategy equilibrium

$$(p \text{ straight} + (1 - p) \text{ swerve}, q \text{ straight} + (1 - q) \text{ swerve}),$$

with $0 < p < 1$ and $0 < q < 1$. From the Fundamental Theorem (Theorem 5.2),

$$p(-2) + (1 - p)(1) = p(-1) + (1 - p)(0),$$
$$q(-2) + (1 - q)(1) = q(-1) + (1 - q)(0).$$

The two equations are the same; this is a consequence of the symmetry of the game. The solution is $(p, q) = (\frac{1}{2}, \frac{1}{2})$. Thus there is a mixed strategy Nash equilibrium in which both players use the same mixed strategy.

Problem 5.12.4 (Rock-Paper-Scissors) has interchangeable players and no pure strategy Nash equilibrium. It has a unique mixed strategy Nash equilibrium in which both players use the same mixed strategy (one-third rock, one-third paper, one-third scissors).

If you did Problem 5.12.8, you saw an example with three interchangeable players in which every pure strategy profile was a Nash equilibrium except the profiles in which all players used the same strategy. There again you found a mixed strategy Nash equilibrium in which every player used the same mixed strategy.

A game in normal form in which the players are interchangeable is called *symmetric*. In such a game, all players, of course, have the same strategy set.

To give a formal definition of a symmetric game, we begin by discussing permutations.

A *permutation* of the set $\{1, 2, \ldots, n\}$ is just a function α from this set to itself that is a bijection. Given such a function α, it can be used to define a reordering of any sequence of length n by the following formula:

$$\alpha \cdot (x_1, x_2, \ldots, x_n) = (x_{\alpha^{-1}(1)}, x_{\alpha^{-1}(2)}, \ldots, x_{\alpha^{-1}(n)}).$$

Note that as is common in mathematics, we identify a sequence of n objects with the ordered n-tuple of those objects.

For example, suppose α is the permutation of the set $\{1, 2, 3, 4\}$ given by

$$\alpha(1) = 2, \quad \alpha(2) = 3, \quad \alpha(3) = 1, \quad \alpha(4) = 4.$$

The inverse is given by

$$\alpha^{-1}(1) = 3, \quad \alpha^{-1}(2) = 1, \quad \alpha^{-1}(3) = 2, \quad \alpha^{-1}(4) = 4.$$

The reordering of sequences defined by α is

$$\alpha \cdot (x_1, x_2, x_3, x_4) = (x_{\alpha^{-1}(1)}, x_{\alpha^{-1}(2)}, x_{\alpha^{-1}(3)}, x_{\alpha^{-1}(4)}) = (x_3, x_1, x_2, x_4).$$

In other words, x_1 moves to location 2, x_2 moves to location 3, x_3 moves to location 1, and x_4 stays in location 4. Thus, when the permutation α is used to reorder a sequence, the meaning of $\alpha(i) = j$ is: move the object in location i to location j.

Here is the result of applying α to various sequences of letters:

$$\alpha \cdot (a, b, c, d) = (c, a, b, d),$$
$$\alpha \cdot (a, a, b, c) = (b, a, a, c),$$
$$\alpha \cdot (a, a, a, b) = (a, a, a, b).$$

Note that if the first three entries are equal, α *fixes* the sequence (i.e., takes the sequence to itself).

Now consider a game with n players, each of whom has the same strategy set S. A mixed strategy profile $(\sigma_1, \ldots, \sigma_n)$ leads to a payoff vector $(\pi_1, \ldots, \pi_n)$. Given a permutation of the set $\{1, 2, \ldots, n\}$, use it to reorder the sequence of strategies in the strategy profile. We obtain a new strategy profile, which leads to a new payoff vector. The game is called *symmetric* if, for every possible reordering of every strategy profile, the new payoff vector is obtained from the previous payoff vector by the same reordering. More formally, the game is symmetric if for every strategy profile $(\sigma_1, \ldots, \sigma_n)$, with associated payoff vector $(\pi_1, \ldots, \pi_n)$, and every permutation α of $\{1, \ldots, n\}$, the payoff vector associated with $\alpha \cdot (\sigma_1, \ldots, \sigma_n)$ is just $\alpha \cdot (\pi_1, \ldots, \pi_n)$.

To check that a game satisfies this definition, it is enough to check that for every reordering of every pure strategy profile, the new payoff vector is obtained from the previous payoff vector by the same reordering. If this is true for all pure strategy profiles, it is also true for all mixed strategy profiles.

As an example, consider the game of Chicken. The strategy profile (straight, swerve) gives the payoff vector $(1, -1)$. Now switch the two strategies, which gives (swerve, straight). The payoff vector is $(-1, 1)$. This vector if obtained by switching the entries in $(1, -1)$. This works for every pure strategy profile, so Chicken is a symmetric game. (There are only two ways to reorder a sequence of two objects: (i) don't do anything, and (ii) switch the objects.)

For a two-player game, the definition of symmetric reduces to the following. Let S denote the set of (pure) strategies available to either player. Then we require

• For all s and t in S, $\pi_1(s, t) = \pi_2(t, s)$.

In other words, if one player uses s and the other uses t, the player who uses s gets a certain payoff. It doesn't matter whether he is Player 1 or Player 2.

The following result was proved by Nash in [11].

Theorem 7.1. *In a symmetric game in which the (pure) strategy set is finite, there is a mixed-strategy Nash equilibrium in which every player uses the same mixed strategy.*

Note that such a strategy profile is fixed by every reordering.

One can take advantage of this theorem by looking for such Nash equilibria instead of more general equilibria.

We note one other fact about symmetric games.

- Suppose a symmetric game has a strategy profile that is a Nash equilibrium, but it is not the case that all players use the same mixed strategy. Then other Nash equilibria can be found by reordering the strategies in the strategy profile.

We have seen this phenomenon in the games of Chicken and Water Pollution, and in Problems 5.12.7 and 5.12.8.

7.2 Reporting a Crime

In 1964 a young woman named Kitty Genovese was murdered outside her home in Queens, New York. According to a *New York Times* article written two weeks later, 38 of her neighbors witnessed the murder, but none of them called the police. While the accuracy of the article has since been questioned (http://en.wikipedia.org/wiki/Kitty_genovese), at the time it horrified the country.

Here is a model that has been proposed for such events. A crime is observed by n people. Each wants the police to be informed but prefers that someone else make the call. Suppose that each person receives a payoff of v as long as at least one person calls the police; if no one calls the police, each person receives a payoff of 0. Each person who calls the police incurs a cost of b. We assume $0 < b < v$.

We view this as a game with n players. Each has two strategies: call the police (c) or don't call the police (d). The total payoffs are:

- If at least one person calls the police: v to each person who does not call, $v - b$ to each person who calls.
- If no one calls the police: 0 to everyone.

You can easily check that there are exactly n pure strategy Nash equilibria. In each of them, exactly one of the n people calls the police.

Motivated by Theorem 7.1, we shall look for a mixed strategy Nash equilibrium $(\sigma, \ldots, \sigma)$ in which all players use the same strictly mixed strategy $\sigma = (1 - p)c + pd$, $0 < p < 1$.

Let's consider Player 1. By the Fundamental Theorem, each of her pure strategies gives her the same expected payoff when Players 2 through n use their mixed strategies:

$$\pi_1(c, \sigma, \ldots, \sigma) = \pi_1(d, \sigma, \ldots, \sigma).$$

Now $\pi_1(c, \sigma, \ldots, \sigma) = v - b$, since the payoff to a player who calls is $v - b$ no matter what the other players do. In contrast,

$$\pi_1(d, \sigma, \ldots, \sigma) = \begin{cases} 0 & \text{if no one else calls,} \\ v & \text{if at least one other person calls.} \end{cases}$$

The probability that no one else calls is p^{n-1}, so the probability that at least one other person calls is $1 - p^{n-1}$. Therefore

$$\pi_1(d, \sigma, \ldots, \sigma) = 0 \cdot p^{n-1} + v \cdot (1 - p^{n-1}) = v(1 - p^{n-1}).$$

Hence

$$v - b = v(1 - p^{n-1}),$$

so

$$p = \left(\frac{b}{v}\right)^{1/(n-1)}.$$

Since $0 < \frac{b}{v} < 1$, p is a number between 0 and 1.

What does this formula mean? Notice first that as $n \to \infty$, $p \to 1$, so $1 - p \to 0$. Thus, as the size of the group increases, each individual's probability of calling the police declines toward 0. However, it is more important to look at the probability that at least one person calls the police. This probability is

$$1 - p^n = 1 - \left(\frac{b}{v}\right)^{n/(n-1)}.$$

As n increases, $n/(n-1) = 1 + 1/(n-1)$ decreases toward 1, so $(b/v)^{n/(n-1)}$ increases toward $\frac{b}{v}$, and so $1 - (b/v)^{n/(n-1)}$ decreases toward $1 - \frac{b}{v}$. Thus, as the size of the group increases, the probability that the police are called *decreases*.

For a large group, the probability that the police are called is approximately $1 - \frac{b}{v}$. Anything that increases b (the perceived cost of calling the police) or decreases v (the value to people of seeing that the police get called) will decrease the likelihood that the police are called.

7.3   Sex Ratio 1

Most organisms that employ sexual reproduction come in two types: male and female. In many species, the percentages of male and female offspring that survive to actually reproduce are very different. Nevertheless, in most species, approximately half of all births are male and half are female. What is the reason for this? This puzzle goes back to Darwin.

One can find an answer by focusing on the number of grandchildren of each female. Suppose a cohort of males and females is about to reproduce. We think of this as a game in which the players are the females, a female's strategy is her fraction of male offspring, and a female's payoff is her number of grandchildren.

7.3.1   Many many females.
There are lots of players! For a first pass at analyzing this situation, imagine that one female has a fraction u of male offspring, $0 \leqslant u \leqslant 1$, and the females as a group have a fraction v of male offspring, $0 \leqslant v \leqslant 1$. We imagine the group is so large that what our one female does has no appreciable effect on v. For each v we will calculate our female's best response set $B(v)$. Motivated by the notion of Nash equilibrium, we ask: for what values of v does the set $B(v)$ include v?

The notation is as follows:

- σ_m = fraction of males that survive to reproduce.
- σ_f = fraction of females that survive to reproduce.
- c = number of offspring per female.
- r = number of offspring per male.
- y = number of females.

Then we have:

	Female 1	All females
Sons	uc	vcy
Daughters	$(1-u)c$	$(1-v)cy$
Surviving sons	$\sigma_m uc$	$\sigma_m vcy$
Surviving daughters	$\sigma_f(1-u)c$	$\sigma_f(1-v)cy$

Let $f(u, v)$ denote the number of grandchildren of Female 1. Then we have

$$f(u, v) = \text{surviving sons} \cdot \text{offspring per son}$$
$$+ \text{surviving daughters} \cdot \text{offspring per daughter}$$
$$= \sigma_m uc \cdot r + \sigma_f(1-u)c \cdot c.$$

The values of σ_m, σ_f, c, and y are given, but r is not. We can calculate r as follows. For the population as a whole,

surviving sons · offspring per son

$\qquad$ = surviving daughters · offspring per daughter,

that is,

$$\sigma_m v c y \cdot r = \sigma_f (1 - v) c y \cdot c.$$

Therefore $r = (\sigma_f(1 - v)/\sigma_m v)c$. Substituting this value into our formula for $f(u, v)$, we obtain

$$f(u, v) = \sigma_m u c \frac{\sigma_f(1 - v)}{\sigma_m v} c + \sigma_f(1 - u)c^2$$
$$= \sigma_f c^2 \left(1 + u\frac{1 - 2v}{v}\right), \quad 0 \leqslant u \leqslant 1,\ 0 < v \leqslant 1.$$

Notice

$$\frac{\partial f}{\partial u}(u, v) = \sigma_f c^2 \frac{1 - 2v}{v} = \begin{cases} + & \text{if } 0 < v < \frac{1}{2}, \\ 0 & \text{if } v = \frac{1}{2}, \\ - & \text{if } v > \frac{1}{2}, \end{cases}$$

Therefore Female 1's best response to v is $u = 1$ if $0 < v < \frac{1}{2}$; any u if $v = \frac{1}{2}$; and $u = 0$ if $v > \frac{1}{2}$. Only in the case $v = \frac{1}{2}$ does Female 1's best response include v.

For Darwin's views on sex ratios, see http://www.ucl.ac.uk/~ucbhdjm/courses/b242/Sex/D71SexRatio.html. For further discussion of sex ratios, see the Wikipedia page http://en.wikipedia.org/wiki/Fisher's_principle.

7.3.2 Not so many females. Suppose there are n females. Female i's strategy is a number u_i, $0 \leqslant u_i \leqslant 1$, that represents the fraction of male offspring she has. Her payoff is her number of grandchildren. Motivated by our work in the previous subsection, we now derive a formula for the payoff.

We continue to use the notation:

- σ_m = fraction of males that survive to reproduce.
- σ_f = fraction of females that survive to reproduce.
- c = number of offspring per female.
- r = number of offspring per male.

Let

$$v = \frac{1}{n}(u_1 + \cdots + u_n).$$

Then we have:

	Female i	All females
Sons	$u_i c$	vcn
Daughters	$(1 - u_i)c$	$(1 - v)cn$
Surviving sons	$\sigma_m u_i c$	$\sigma_m vcn$
Surviving daughters	$\sigma_f(1 - u_i)c$	$\sigma_f(1 - v)cn$

The formula for r is unchanged. We therefore have

$$\pi_i(u_1, \ldots, u_n) = \sigma_f c^2 \left(1 + u_i \frac{1 - 2v}{v}\right).$$

We will look for a Nash equilibrium at which $0 < u_i < 1$ for every i. Then for every i, we must have $(\partial \pi_i / \partial u_i)(u_1, \ldots, u_n) = 0$, that is,

$$0 = \frac{\partial \pi_i}{\partial u_i} = \sigma_f c^2 \left(1 \cdot \frac{1 - 2v}{v} + u_i \frac{d}{dv}\left(\frac{1 - 2v}{v}\right) \frac{\partial v}{\partial u_i}\right)$$

$$= \sigma_f c^2 \left(\frac{1 - 2v}{v} - u_i \frac{1}{v^2} \frac{1}{n}\right) = \sigma_f c^2 \frac{nv - 2nv^2 - u_i}{nv^2}.$$

Hence, for every i, $u_i = nv - 2nv^2$. Therefore all u_i are equal; denote their common value by u. Then $v = u$, so

$$0 = \frac{nu - 2nu^2 - u}{nu^2} = \frac{n - 2nu - 1}{nu}.$$

Therefore

$$u = \frac{n - 1}{2n} = \frac{1}{2} - \frac{1}{2n}.$$

For n large, u is very close to $\frac{1}{2}$.

This game is symmetric, and in the Nash equilibrium we have found, every player uses the same strategy. Could we have assumed from the beginning that, because of Theorem 7.1, there must be a Nash equilibrium in which all the u_i are equal? No; Theorem 7.1 does not apply. This is not a game with a finite pure strategy set. Instead, the ith player has an interval $0 \leqslant u_i \leqslant 1$ of pure strategies.

7.4 Other symmetries of games

Nash actually proved a result about symmetries of games that is more general than Theorem 7.1. In this section we describe two more situations to which Nash's general result applies.

7.4.1 More about permutations. We begin by studying permutations a little more. Here are some important facts about permutations:

(1) Since a permutation is just a function from the set $\{1, 2, \ldots, n\}$ to itself, two permutations of $\{1, 2, \ldots, n\}$ can be composed. The result is a new permutation of $\{1, 2, \ldots, n\}$. In other words, the permutation α followed by the permutation β gives the permutation $\beta \circ \alpha$.

(2) The inverse of a permutation α of $\{1, 2, \ldots, n\}$ is another permutation α^{-1} of $\{1, 2, \ldots, n\}$ that undoes it. We have $\alpha^{-1} \circ \alpha = \alpha \circ \alpha^{-1} = \iota$, where ι denotes the identity pemutation: $\iota(i) = i$ for every i.

The collection of all permutations of $\{1, 2, \ldots, n\}$ is called S_n, the *symmetric group of order n*. A *subgroup* of S_n is a nonempty subset $\mathcal{H}$ of S_n that is closed under composition and taking inverses:

(1) if $\alpha \in \mathcal{H}$ and $\beta \in \mathcal{H}$, then $\beta \circ \alpha \in \mathcal{H}$;
(2) if $\alpha \in \mathcal{H}$, then $\alpha^{-1} \in \mathcal{H}$.

Since $\alpha^{-1} \circ \alpha = \iota$, ι is in every subgroup of S_n.

The smallest subgroup of S_n is the subgroup consisting of just ι. Other subgroups of S_n can have as few as two elements. For example, let α be the permutation of $\{1, 2, \ldots, n\}$ given by

$$\alpha(1) = n, \ \alpha(2) = n - 1, \ \ldots, \ \alpha(n - 1) = 2, \ \alpha(n) = 1.$$

Then α is its own inverse. Therefore the set $\mathcal{H} = \{\iota, \alpha\}$ is a subgroup of S_n.

7.4.2 Symmetries of the players. Let G be an n-player game in normal form. The players are denoted $1, 2, \ldots, n$, so the set $\{1, 2, \ldots, n\}$ is just the set of players. We do not require that all players have the same strategy set, but there may be subsets of the players with the same strategy set. More precisely, we assume that the players can be divided into k *types*; all players of the same type have the same strategy set. In Section 7.1 we considered the case where there was just one type.

We say that a permutation α of the set $\{1, 2, \ldots, n\}$ of players *respects the types in the game G* if for every $i = 1, \ldots, n$, $\alpha(i)$ is a player of the same type as Player i. The subset of S_n consisting permutations that respect the types in G is a subgroup of S_n that we denote S_G.

As in Section 7.1, we use permutations of $\{1, 2, \ldots, n\}$ to reorder mixed strategy profiles $(\sigma_1, \ldots, \sigma_n)$. However, the only permutations of $\{1, 2, \ldots, n\}$ that it makes sense to use are those in S_G. For example, if α is the permutation of $\{1, 2, \ldots, n\}$ with $\alpha(1) = 2$, $\alpha(2) = 1$, and $\alpha(i) = i$ for $i > 2$, then

$$\alpha \cdot (\sigma_1, \sigma_2, \sigma_3, \ldots, \sigma_n) = (\sigma_2, \sigma_1, \sigma_3, \ldots, \sigma_n).$$

This makes sense if σ_1 is a mixed strategy of Player 2 and σ_2 is a mixed strategy of Player 1. This is true when Players 1 and 2 are of the same type, which is true when $\alpha \in S_G$.

Let $\mathcal{H}$ be a subgroup of S_G. Suppose that when we use any permutation in $\mathcal{H}$ to reorder any mixed strategy profile $(\sigma_1, \ldots, \sigma_n)$, the payoff vector for the new strategy profile is obtained by applying the same permutation to reorder the payoff vector for the old strategy profile. Then we say that the game is *invariant under the symmetry group $\mathcal{H}$ of its players*.

Theorem 7.2. *Suppose a game is invariant under a symmetry group $\mathcal{H}$ of its players. Then it has a Nash equilibrium $(s_1^*, \ldots, s_n^*)$ that is fixed by every permutation in $\mathcal{H}$. In other words, if we use any permutation in $\mathcal{H}$ to reorder $(s_1^*, \ldots, s_n^*)$, the result is again $(s_1^*, \ldots, s_n^*)$.*

For example, suppose a game has seven players who come in two types. Players 1 through 4 are of the first type, and Players 5 through 7 are of the second type. Within a type, the players are interchangeable. Let's translate this informal description into the language of symmetry groups. S_G is the subgroup of S_7 consisting of permutations of $\{1, \ldots, 7\}$ that can be described as a permutation of $\{1, 2, 3, 4\}$ followed by a permutation of $\{5, 6, 7\}$. When we say that players of each type are interchangeable, we mean that the game is invariant under the full symmetry group S_G of its players. Then Theorem 7.2 implies that it has a Nash equilibrium that is fixed by every permutation in S_G. The only strategy profiles that are fixed by every permutation in S_G are those for which the first four players all use the same strategy, and the last three players all use the same strategy. Therefore there is a Nash equilibrium of this type.

7.4.3 Symmetries of the strategies.
In this subsection we discuss permuting individual players' strategies without permuting the players. We consider an n-player game in normal form in which the ith player has a strategy set S_i with k_i pure strategies: $S_i = \{s_{i1}, s_{i2}, \ldots s_{ik_i}\}$. We let ℓ denote the (long) sequence of strategies in S_1 followed by strategies in S_2 followed by …followed by strategies in S_n; in the sequence ℓ, the strategies in S_i are written in the order $s_{i1}, s_{i2}, \ldots s_{ik_i}$. Then ℓ is a sequence of $k = k_1 + k_2 + \cdots + k_n$ strategies. Let $\mathcal{H}$ be a subgroup of S_k consisting of permutations of the set $\{1, 2, \ldots, k\}$ that can be expressed as a permutation of the first k_1 numbers followed by a permutation of the next k_2 numbers followed by …followed by a permutation of of the last k_n numbers. If $\alpha \in \mathcal{H}$ and we use α to reorder the sequence ℓ, then α will move each strategy of Player i to the location of another strategy of Player i.

Each such permutation α can be used to define a function $\tilde{\alpha}$ from the set of pure strategy profiles $S_1 \times S_2 \times \cdots \times S_n$ to itself as follows. Let $(s_1, s_2, \ldots, s_n)$

be a profile of pure strategies. Suppose that when α is used to reorder the sequence ℓ, α takes s_1 to the location of another strategy s_1' in S_1, s_2 to the location of another strategy s_2' in $S_2, \ldots, s_n$ to the location of another strategy s_n' in S_n. Then

$$\tilde{\alpha}(s_1, s_2, \ldots, s_n) = (s_1', s_2', \ldots, s_n').$$

In fact α also induces a mapping of mixed strategy profiles, which we will still call $\tilde{\alpha}$. For example, suppose that when α is used to reorder the sequence ℓ, α puts s_{11} in the s_{13} location, and puts s_{12} in the s_{14} location. Consider a mixed strategy profile $(\sigma_1, \ldots, \sigma_n)$ with $\sigma_1 = p_1 s_{11} + p_2 s_{12}$. Then

$$\tilde{\alpha}(\sigma_1, \ldots, \sigma_n) = (\sigma_1', \ldots, \sigma_n') \quad \text{where } \sigma_1' = p_1 s_{13} + p_2 s_{14}.$$

Suppose that every permutation α in $\mathcal{H}$ induces a mapping $\tilde{\alpha}$ of pure strategy profiles such that each pure strategy profile $(s_1, s_2, \ldots, s_n)$ has the same payoff vector as $\tilde{\alpha}(s_1, s_2, \ldots, s_n)$. Then we say that the game is *invariant under the symmetry group $\mathcal{H}$ of its strategies*. In this case it is also true that every mixed strategy profile $(\sigma_1, \ldots, \sigma_n)$ has the same payoff vector as $\tilde{\alpha}(\sigma_1, \ldots, \sigma_n)$.

Theorem 7.3. *Suppose a game is invariant under a symmetry group $\mathcal{H}$ of its strategies. Then it has a mixed strategy Nash equilibrium $(\sigma_1^*, \ldots, \sigma_n^*)$ that is fixed by every permutation in $\mathcal{H}$. In other words, for any permutation α in $\mathcal{H}$, $\tilde{\alpha}(\sigma_1^*, \ldots, \sigma_n^*) = (\sigma_1^*, \ldots, \sigma_n^*)$.*

In examples, this theorem implies that a game has a Nash equilibrium in which certain probabilities are equal.

7.4.4 Colonel Blotto revisited. To see what Theorem 7.3 means, consider the game of Colonel Blotto vs. the People's Militia from Section 5.7. The payoff matrix is reproduced in Table 7.1.

Table 7.1.

		People's Militia			
		30	21	12	03
	40	$(4, -4)$	$(2, -2)$	$(1, -1)$	$(0, 0)$
	31	$(1, -1)$	$(3, -3)$	$(0, 0)$	$(-1, 1)$
Col. Blotto	22	$(-2, 2)$	$(2, -2)$	$(2, -2)$	$(-2, 2)$
	13	$(-1, 1)$	$(0, 0)$	$(3, -3)$	$(1, -1)$
	04	$(0, 0)$	$(1, -1)$	$(2, -2)$	$(4, -4)$

You can see a symmetry in this payoff matrix: if you flip the matrix of payoff vectors across a horizontal line through the middle, then flip again across a vertical line though the middle, the matrix of payoff vectors is unchanged. In other words, if we "reverse" both players' strategies, the payoffs don't change.

To express this symmetry in Nash's language, let ℓ denote the sequence of strategies in Col. Blotto's strategy set $S_1 = \{40, 31, 22, 13, 04\}$, followed by strategies in the People's Militia's strategy set $S_2 = \{30, 21, 12, 03\}$, written in that order:

$$\ell = (40, 31, 22, 13, 04; 30, 21, 12, 03).$$

The semicolon is there just to divide the two players' strategies. Now consider the permutation α of the set $\{1, \ldots, 9\}$ that reverses the numbers $1, \ldots, 5$ and reverses the numbers $6, \ldots, 9$:

$$\alpha(1) = 5, \quad \alpha(2) = 4, \quad \alpha(3) = 3,$$
$$\alpha(4) = 2, \quad \alpha(5) = 1, \quad \alpha(6) = 9,$$
$$\alpha(7) = 8, \quad \alpha(8) = 7, \quad \alpha(9) = 6.$$

α is its own inverse, so $\mathcal{H} = \{\iota, \alpha\}$ is a subgroup of S_9. Moreover, when α is used to reorder the sequence ℓ, α takes each strategy in S_1 to the location of another strategy in S_1, and takes each strategy in S_2 to the location of another strategy in S_2. In fact,

$$\alpha \cdot \ell = \alpha \cdot (40, 31, 22, 13, 04; 30, 21, 12, 03) = (04, 13, 22, 31, 40; 03, 12, 21, 30).$$

This shows the position to which α takes each strategy. We use it to define a function $\tilde{\alpha}$ from $S_1 \times S_2$ to itself. For example, to calculate $\tilde{\alpha}(40, 12)$, we note that α takes 40 to the original position of 04 and takes 12 to the original position of 21; therefore $\tilde{\alpha}(40, 12) = (04, 21)$.

To check that this game is invariant under the symmetry group $\mathcal{H}$ of its strategies, we must check that the payoff vectors associated to $(40, 12)$ and $\tilde{\alpha}(40, 12) = (04, 21)$ are the same. You can easily check that this is correct. Of course, we must also check 19 similar equations.

The checks all work because of the visual symmetry in the payoff matrix that we noted earlier. One way to express this is to write the payoff matrix with the strategies written as shown in Table 7.2. The checks we have to do amount to checking that this payoff matrix is identical to the first. You can see that they are.

Since our game is invariant under the symmetry group $\mathcal{H} = \{i, \alpha\}$ of its strategies, Theorem 7.3 says that it has a mixed strategy Nash equilibrium

$$(\sigma_1, \sigma_2) = (p_1 40 + p_2 31 + p_3 22 + p_4 13 + p_5 04, q_1 30 + q_2 21 + q_3 12 + q_4 03)$$

Table 7.2.

		People's Militia			
		03	12	21	30
	04	$(4, -4)$	$(2, -2)$	$(1, -1)$	$(0, 0)$
	13	$(1, -1)$	$(3, -3)$	$(0, 0)$	$(-1, 1)$
Col. Blotto	22	$(-2, 2)$	$(2, -2)$	$(2, -2)$	$(-2, 2)$
	31	$(-1, 1)$	$(0, 0)$	$(3, -3)$	$(1, -1)$
	40	$(0, 0)$	$(1, -1)$	$(2, -2)$	$(4, -4)$

that is fixed by the extension of $\tilde{\alpha}$ to mixed strategy profiles. Now

$$\tilde{\alpha}(\sigma_1, \sigma_2)$$
$$= \tilde{\alpha}(p_1 40 + p_2 31 + p_3 22 + p_4 13 + p_5 04, q_1 30 + q_2 21 + q_3 12 + q_4 03)$$
$$= (p_1 04 + p_2 13 + p_3 22 + p_4 31 + p_5 40, q_1 03 + q_2 12 + q_3 21 + q_4 30).$$

For (σ_1, σ_2) to be fixed by $\tilde{\alpha}$, we must have $p_1 = p_5$, $p_2 = p_4$, $q_1 = q_4$, and $q_2 = q_3$. Thus Theorem 7.3 tells us that there is a Nash equilibrium of the form

$$(\sigma_1, \sigma_2) = (a40 + b31 + c22 + b13 + a04, d30 + e21 + e12 + d03). \quad (7.1)$$

We found such a Nash equilibrium in Section 5.7 after considerable work:

$$a = \tfrac{4}{9}, \qquad b = 0, \qquad c = \tfrac{1}{9}, \qquad d = \tfrac{1}{18}, \qquad e = \tfrac{4}{9}.$$

You could find this Nash equilibrium more easily by assuming there is one of the form (7.1) with all strategies active except Col. Blotto's 31 and 13, so that $b = 0$. Write down the usual equations. Many will be redundant and can be dropped. You should wind up with just two equations to determine a and c, and two to determine d and e. There is still an inequality check to do at the end (actually two, but one is redundant).

7.5 Problems

7.5.1 The Princeton Bar. This problem is based on a scene in the movie *A Beautiful Mind* about the life of John Nash. n men walk into a bar. In the bar is one extremely attractive woman and many attractive women. Each man has two possible pure strategies:

- s: Approach one of the attractive women. (The safe strategy.)
- r: Approach the extremely attractive woman. (The risky strategy.)

The payoffs are:

- $a > 0$ to each man who uses strategy s. (There are many attractive women in the bar; the strategy of approaching one of them will succeed.)
- If there is a unique man who uses strategy r, his payoff is $b > a$. If more than one man uses strategy r, they all have payoff 0. (The extremely attractive woman doesn't enjoy being pestered and leaves.)

(1) Find all pure strategy Nash equilibria of this n-player game.
(2) Find a mixed strategy Nash equilibrium in which all n men use the same mixed strategy $ps + (1 - p)r$.
(3) In the Nash equilibrium of part (2), for large n, what is the approximate probability that at least one man approaches the extremely attractive woman?

By the way, the movie *A Beautiful Mind* wrongly implies that it is a Nash equilibrium for none of the men to approach the extremely attractive woman.

7.5.2 The Sneaky Cat 1. A cat is considering sneaking up on a bird. The cat has two strategies: hightail it out of there and look for something else to eat (h), or sneak up on the bird (s).

The bird has two strategies: trust that the cat will not try to sneak up on it (t), or watch out for the cat (w). If the cat stalks the bird and the bird does not watch out, the cat will get the bird.

Let 1 be the value to the cat of eating the bird and the cost to the bird of being eaten. Let r be the cost to the cat of stalking the bird, and let c be the cost to the bird of watching out for the cat. We get the following payoffs:

		Bird	
		t	w
Cat	h	$(0,0)$	$(0,-c)$
	s	$(1-r,-1)$	$(-r,-c)$

We assume $0 < r < 1$ and $c > 0$. Notice that 0 is the bird's best payoff.

(1) Show that if $c > 1$, (s,t) is a Nash equilibrium.
(2) Show that if $0 < c < 1$, there are no pure strategy Nash equilibria.
(3) For $0 < c < 1$, find a mixed strategy Nash equilibrium.

7.5.3 The Sneaky Cat 2. Now suppose there is one cat and a flock of n birds. The cat still has the strategies h and s. Each bird has the strategies t and w. The payoffs in this $(n + 1)$-player game are as follows:

- Cat uses h: 0 to the cat, 0 to each bird that uses t, $-c$ to each bird that uses w.
- Cat uses s, all birds use t: $1 - r$ to the cat (the cat gets a bird if no birds watch out, but the other birds fly off), $-\frac{1}{n}$ to each bird (each bird has probability $\frac{1}{n}$ of being the unlucky one).
- Cat uses s, at least one bird uses w: $-r$ to the cat (if at least one bird is watching out, all birds fly off, and the cat goes hungry), 0 to each bird that uses t, $-c$ to each bird that uses w.

(1) Assume $0 < c < \frac{1}{n}$. Show that there are no pure strategy Nash equilibria. (Consider the following cases: (i) cat uses h, all birds use t; (ii) cat uses h, at least one bird uses w; (iii) cat uses s, all birds use t; (iv) cat uses s, at least one bird uses w.)

(2) Assume $\frac{1}{n} r^{(n-1)/n} < c < \frac{1}{n}$. Find a Nash equilibrium in which the cat uses s, and all birds use the same mixed strategy $\tau = qt + (1 - q)w$, $0 < q < 1$. (For $(s, \tau, \tau, \ldots, \tau)$ to be a Nash equilibrium, we need: (i) if bird 1, for example, uses instead one of the pure strategies t or w, his expected payoff is the same; (ii) if the cat uses instead the pure strategy h, his payoff does not go up. Use (i) to find q, then check (ii).)

(3) For $0 < c < \frac{1}{n} r^{(n-1)/n}$, find a Nash equilibrium in which the cat uses a mixed strategy $\sigma = ph + (1 - p)s$, $0 < p < 1$, and all birds use the same mixed strategy $\tau = qt + (1 - q)w$, $0 < q < 1$.

The fact that there is a Nash equilibrium in which all birds use the same strategy is a consequence of Theorem 7.2.

7.5.4 Colonel Blotto continued.
Use the method proposed in the last paragraph of Subsection 7.4.4 to find the symmetric Nash equilibrium of the Colonel Blotto game.

7.5.5 Rock-Paper-Scissors 2.
In the Nash equilibrium of Rock-Paper-Scissors (Problem 5.12.4), each player uses each strategy with probability one-third. In this problem we will use Nash's ideas about symmetries of strategies (Subsection 7.4.3) to explain this fact. We will follow the approach of Subsection 7.4.4 to Colonel Blotto vs. the People's Militia.

In Rock-Paper-Scissors, let ℓ denote the sequence of strategies in Player 1's strategy set $S_1 = \{r, p, s\}$, followed by strategies in Player 2's strategy set $S_2 = \{r, p, s\}$, written in that order: $\ell = (r, p, s; r, p, s)$. The semicolon is just there to divide the two players' strategies. Now consider the permutation α

of the set $\{1, 2, 3, 4, 5, 6\}$ given by

$$\alpha(1) = 2, \qquad \alpha(2) = 3, \qquad \alpha(3) = 1,$$
$$\alpha(4) = 5, \qquad \alpha(5) = 6, \qquad \alpha(6) = 4.$$

(1) Calculate $\alpha^2 = \alpha \circ \alpha$.

(2) Show that α^2 is the inverse of α.

Therefore $\mathcal{H} = \{\iota, \alpha, \alpha^2\}$ is a subgroup of S_6. We can think of α as a permutation of $\{1, 2, 3\}$ (the positions of Player 1's strategies in ℓ) followed by a permutation of $\{4, 5, 6\}$ (the positions of Player 2's strategies in ℓ), and of course we can think of α^2 the same way.

We have

$$\alpha \cdot \ell = \alpha \cdot (r, p, s; r, p, s) = (s, r, p; s, r, p),$$
$$\alpha^2 \cdot \ell = \alpha \cdot (\alpha \cdot \ell) = \alpha \cdot (s, r, p; s, r, p) = (p, s, r; p, s, r).$$

These formulas show the position to which α and α^2 take each strategy. We use them to define functions $\tilde{\alpha}$ and $\tilde{\alpha}^2$ from $S_1 \times S_2$ to itself. For example, to calculate $\tilde{\alpha}(r, s)$, we note that α takes the first r to the original position of the first p, and takes the second s to the original position of the second r; therefore $\tilde{\alpha}(r, s) = (p, r)$.

(3) Calculate $\tilde{\alpha}$ of all other pure strategy profiles. (There are nine pure strategy profiles in all.)

To check that this game is invariant under the symmetry group $\mathcal{H}$ of its strategies, we must check, for example, that the payoff vectors associated to (r, s) and $\tilde{\alpha}(r, s) = (p, r)$ are the same. Using the payoff matrix of Problem 5.12.4, you can easily check that this is correct: both are $(1, -1)$. There are eight similar equations that must be checked for $\tilde{\alpha}$.

(4) Here is a quick way to check all nine equations for $\tilde{\alpha}$. Look at the payoff matrix in Problem 5.12.4. Rewrite the payoff matix with the strategies in the new order given by $\alpha \cdot \ell$ (i.e., list the strategies in the order s, r, p for both players. Check that the new payoff matrix is the same as the old).

One could do a similar check for $\tilde{\alpha}^2$; we won't do it. Actually, the fact that the nine equations for $\tilde{\alpha}^2$ hold is a consequence of the fact that the nine equations for $\tilde{\alpha}$ hold. Do you see why?

Thus Rock-Paper-Scissors is invariant under the symmetry group $\mathcal{H} = \{\iota, \alpha, \alpha^2\}$ of its strategies. Then Theorem 7.3 says that it has a mixed strategy Nash equilibrium

$$(\sigma_1, \sigma_2) = (p_1 r + p_2 p + p_3 s, q_1 r + q_2 p + q_3 s)$$

that is fixed by the extension of $\tilde{\alpha}$ to mixed strategy profiles. Now

$$\tilde{\alpha}(\sigma_1, \sigma_2) = \tilde{\alpha}(p_1r + p_2p + p_3s, q_1r + q_2p + q_3s)$$
$$= (p_1p + p_2s + p_3r, q_1p + q_2s + q_3r).$$

(5) Show that the only strategy profile (σ_1, σ_2) that is fixed by $\tilde{\alpha}$ has $p_1 = p_2 = p_3 = q_1 = q_2 = q_3 = \frac{1}{3}$.

(6) Theorem 7.3 implies that this strategy profile must be a Nash equilibrium; no further checks need be done. Explain.

Chapter 8

Alternatives to the Nash equilibrium

In this chapter we consider three points of view that lead to alternatives to the Nash equilibrium. The notion of a *correlated equilibrium* relies on social rules to direct players to appropriate actions. *Epistemic game theory* formalizes how players' beliefs about one another lead them to actions. Finally, the notion of *evolutionary stability* was introduced by biologists to describe strategy profiles in populations that resist invasion by organisms using other strategies. The invaders may come from elsewhere, or they may arise within the population by mutation.

8.1 Correlated equilibrium

Two drivers arrive at an intersection, Driver 1 from the south and Driver 2 from the east. Each has two strategies: go and stop. If one goes and one stops, we will take the payoffs to be -1 to the driver who stops for time lost waiting, and 0 to the driver who goes, since he does not lose any time. If both stop, we take the expected payoff to each to be -2; one will eventually go first. If both go, there is the possibility of a serious accident; we take the expected payoff to each to be -4. We obtain the following payoff matrix:

		Driver 2	
		go	stop
Driver 1	go	$(-4, -4)$	$(0, -1)$
	stop	$(-1, 0)$	$(-2, -2)$

This game is a variant of Chicken: the best response to the other player's strategy is the opposite strategy. There are two pure-strategy Nash equilibria, (go, stop) and (stop, go). The first is better for Player 1, the second is better for Player 2.

In practice, this problem is solved by a traffic light or a driving rule. A traffic light signals green to one driver and red to the other, telling each which to do.

A driving rule says, for example, that the driver on the right has the right-of-way. If this is the rule, Driver 2 goes. Either rule has the following property: if the other driver follows the rule, you get your best payoff by also following the rule.

To generalize this example, we consider a *game in normal form with public signals*. This is a game in normal form with n players and, in addition, a set of public signals ω_k. Each public signal ω_k occurs with probability α_k; each α_k is positive, and the sum of the α_ks is 1. Each player observes the signal ω_k and chooses one of his mixed strategies in response. Thus the jth player has a *response function* from the set of signals to his set of mixed strategies. We denote the strategy chosen by the jth player in response to the signal ω_k by $\sigma_j(\omega_k)$. An ordered n-tuple of response functions is said to be a *correlated equilibrium* if for each signal ω_k, the strategy profile $(\sigma_1(\omega_k), \ldots, \sigma_n(\omega_k))$ is a Nash equilibrium.

Traffic lights and driving rules are both covered by this framework. The traffic light has two signals: ω_1, which is green north-south and red east-west, and ω_2, which is red north-south and green east-west. For the correlated equilibrium that obtains in practice, Driver 1 chooses go in response to ω_1 and stop in response to ω_2; Driver 2 chooses the opposite. With the traffic rule described above, there is just one signal, namely, the rule. Driver 1 chooses stop in response, and Driver 2 chooses go.

Another game that has a nice solution when public signals are added is Battle of the Sexes (Subsection 3.2.4). For example, suppose Alice and Bob have an agreement that weekdays are for concerts and weekends are for wrestling. Then the day of the week serves as a public signal that allows them to coordinate their behavior.

In general, some players may be unable to distinguish among some signals. If, for example, Player 1 is unable to distinguish between signals ω_1 and ω_2, then his response function is required to assign the same response to both signals. Instead of requiring that the players' responses to each signal form a Nash equilibrium, we require that each player's response to each signal be a best response to his best estimate of the other players' responses. We denote Player i's best estimate of Player j's response when the signal is ω_k by $\tau_{ij}(\omega_k)$. Of course, if Player i is unable to distinguish between the signals ω_k and $\omega_{k'}$, we must have $\tau_{ij}(\omega_k) = \tau_{ij}(\omega_{k'})$ for every j.

A collection of signals that Player i is unable to distinguish is called an *information set* of Player i. For example, suppose that Player 1 has an information set consisting of signals ω_1 and ω_2. If signal ω_1 is sent, then Player 1 does not know whether signal ω_1 or signal ω_2 was sent, so he does not know whether the other players have observed signal ω_1 or signal ω_2. From Player 1's point of view, another Player j has observed signal ω_1 with probability $\alpha_1/(\alpha_1 + \alpha_2)$ and signal ω_2 with probability $\alpha_2/(\alpha_1 + \alpha_2)$. Thus, from Player 1's point of view, Player j will use the mixed strategy $\sigma_j(\omega_1)$ with probability $\alpha_1/(\alpha_1 + \alpha_2)$, and $\sigma_j(\omega_2)$ with probability $\alpha_2/(\alpha_1 + \alpha_2)$. Thus Player 1's best estimate of Player j's response when the signal is ω_1 is

$$\tau_{1j}(\omega_1) = \tau_{1j}(\omega_2) = \frac{\alpha_1}{\alpha_1 + \alpha_2}\sigma_j(\omega_1) + \frac{\alpha_2}{\alpha_1 + \alpha_2}\sigma_j(\omega_2).$$

In a correlated equilibrium with signal ω_1, Player 1's strategy $\sigma_1(\omega_1)$ is required to be a best response to $(\tau_{12}(\omega_1),\ldots,\tau_{1n}(\omega_1))$, and analogously for the other players.

Response functions to signals are sometimes thought of as *social norms*. In a correlated equilibrium, if the other players follow the social norm, the best you can do is also to follow it.

A difficulty with the Nash equilibrium is deciding what strategy to use when there are several equally plausible Nash equilibria. Public signals and correlated equilibria are a solution to this problem. Of course, another solution is mixed strategies. However, correlated equilibria are often much better for both players. In the Driver Game, for example, if the traffic light gives each signal with probability $\frac{1}{2}$, then the expected payoff to each player from the given correlated equilibrium is $-\frac{1}{2}$. In contrast, in the mixed-strategy Nash equilibrium, each player goes with probability $\frac{2}{5}$ and stops with probability $\frac{3}{5}$. The expected payoff to each player is $-\frac{8}{5}$.

8.2 Epistemic game theory

Epistemic game theory is the branch of game theory that emphasizes the beliefs that players have about their opponents. For a two-player game, one way to formalize players' beliefs and their consequences is as follows.

Consider a two-player game in normal form, in which Player 1's strategy set is $S = \{s_1,\ldots,s_n\}$ and Player 2's strategy set is $T = \{t_1,\ldots,t_m\}$. A *belief* of Player 1 about Player 2 is a finite list of mixed strategies τ_j that Player 2 might play, together with an assignment of probabilities $q_j > 0$ to each τ_j on the list. Similarly, a belief of Player 2 about Player 1 is a finite list of mixed strategies σ_i that Player 1 might play, together with an assignment of probabilities $p_i > 0$ to each σ_i on the list. Given the players' beliefs, each player chooses a mixed strategy that maximizes his own expected payoff given his own beliefs. Thus Player 1 selects a mixed strategy σ^* such that

$$\sum q_j \pi_1(\sigma^*, \tau_j) \geqslant \sum q_j \pi_1(\sigma, \tau_j) \quad \text{for all mixed strategies } \sigma \text{ of Player 1,}$$

and Player 2 selects a mixed strategy τ^* such that

$$\sum p_i \pi_2(\sigma_i, \tau^*) \geqslant \sum p_i \pi_2(\sigma_i, \tau) \quad \text{for all mixed strategies } \tau \text{ of Player 2.}$$

Equivalently, Player 1 chooses a best response σ^* to the mixed strategy $\sum q_j \tau_j$ of Player 2, and Player 2 chooses a best response to the mixed strategy $\sum p_i \sigma_i$ of Player 1. The payoffs are then those that result from the strategy profile (σ^*, τ^*).

For example, consider Rosenthal's Centipede Game from Section 1.8. A strategy for Player 1 (Mutt) is a plan, for each node that is labeled with his name, whether to cooperate or defect should that node be reached. For determining payoffs, the only relevant fact about Mutt's strategy is the first node at which he plans to defect. Thus we shall let s_i denote the strategy, first defect at Mutt's ith node, where $i = 1, \ldots, 98$. Similarly, for Player 2 (Jeff), we shall let t_j denote the strategy, first defect at Jeff's jth node, where $j = 1, \ldots, 98$.

As discussed in Section 1.13, in Rosenthal's Centipede Game, each player's belief about the other is related to how many steps of backward induction he expects the other to do. Suppose, for example, that each player expects the other to cooperate for 95 turns before defecting, that is, Player 1 believes Player 2 will use his strategy t_{96} with probability 1, and Player 2 believes Player 1 will use his strategy s_{96} with probability 1. The best response to t_{96} is s_{96}, and the best response to s_{96} is t_{95}. Thus the strategy profile that is played is (s_{96}, t_{95}). Player 1's belief turns out to be wrong and Player 2's to be right. The game ends when Jeff defects on his 95th turn (Mutt was planning to defect on his next turn but doesn't get to). The payoffs are 95 to Mutt and 98 to Jeff.

For another example, consider the Traveler's Dilemma (Problem 2.14.6). Let $s_i = t_i =$ report expenses of $i + 1$ dollars, $i = 1, 2, 3, 4$. Suppose each salesman expects the other to report expenses of \$3 with probability .4 and \$4 with probability .6. Then Salesman 1 must choose his mixed strategy $\sigma = \sum p_i s_i$ to maximize the expression

$$
\begin{aligned}
\pi_1(\sigma, .4t_2 + .6t_3) &= \pi_1(p_1 s_1 + p_2 s_2 + p_3 s_3 + p_4 s_4, .4t_2 + .6t_3) \\
&= p_1(.4 \cdot 4 + .6 \cdot 4) + p_2(.4 \cdot 3 + .6 \cdot 5) \\
&\quad + p_3(.4 \cdot 1 + .6 \cdot 4) + p_4(.4 \cdot 1 + .6 \cdot 2) \\
&= 4p_1 + 4.2p_2 + 2.8p_3 + 2p_4
\end{aligned}
$$

subject to the constraints $p_i \geqslant 0$ and $\sum p_i = 1$. Because of the constraints, this expression is maximum when $p_2 = 1$ and the other $p_i = 0$. Thus Salesman 1 uses his strategy s_2 (i.e., he reports expenses of \$3). Salesman 2 does the same. Payoffs are \$3 to each salesman.

8.3 Evolutionary stability

Evolutionary game theory focuses on *populations* that repeatedly play games, rather than on individuals.

We consider a symmetric two-player game G in normal form in which the strategy set S is finite: $S = \{s_1, \ldots, s_n\}$. We recall that symmetry means

$$\pi_2(s_i, s_j) = \pi_1(s_j, s_i).$$

In an *evolutionary game* there is a population of many individuals who play the game G with one another. An individual uses a mixed strategy $\tau = \sum q_i s_i$ with (of course) all $q_i \geqslant 0$ and $\sum q_i = 1$. We say that the individual is of *type* τ. If τ is a pure strategy, say, $\tau = s_i$, we say that the individual is of type i.

The population taken as a whole uses strategy s_1 with probability $p_1, \ldots,$ strategy s_n with probability p_n; all $p_i \geqslant 0$ and $\sum p_i = 1$. We say that the *state* of the population is σ and write $\sigma = \sum p_i s_i$.

If an individual of type i plays this game against a random individual from a population of type σ, her expected payoff is just the expected payoff to an individual who uses the pure strategy s_i against another individual using the mixed strategy σ:

$$\pi_1(s_i, \sigma) = \sum_{j=1}^{n} \pi_1(s_i, s_j) p_j.$$

If an individual of type τ plays this game against a random individual from a population of type σ, her expected payoff is just the expected payoff to an individual who uses the mixed strategy τ against another individual using the mixed strategy σ:

$$\pi_1(\tau, \sigma) = \sum_{i=1}^{n} q_i \pi_1(s_i, \sigma) = \sum_{i,j=1}^{n} q_i \pi_1(s_i, s_j) p_j. \tag{8.1}$$

Let σ_1 and σ_2 be population states, $\sigma_1 = \sum p_{1i} s_i$, $\sigma_2 = \sum p_{2i} s_i$, and let $0 < \epsilon < 1$. Then we can define a new population state

$$(1 - \epsilon)\sigma_1 + \epsilon\sigma_2 = \sum ((1 - \epsilon)p_{1i} + \epsilon p_{2i}) s_i.$$

If τ is a mixed strategy, one can easily show from (8.1) that

$$\pi_1(\tau, (1 - \epsilon)\sigma_1 + \epsilon\sigma_2) = (1 - \epsilon)\pi_1(\tau, \sigma_1) + \epsilon\pi_1(\tau, \sigma_2). \tag{8.2}$$

Suppose, in a population with state σ, we replace a fraction ϵ of the population, $0 < \epsilon < 1$, with individuals of type τ. The new population state is $(1 - \epsilon)\sigma + \epsilon\tau$.

We say that a population state σ is *evolutionarily stable* if for each $\tau \neq \sigma$ there is a number $\epsilon_0 > 0$ such that if $0 < \epsilon < \epsilon_0$ then

$$\pi_1(\sigma, (1 - \epsilon)\sigma + \epsilon\tau) > \pi_1(\tau, (1 - \epsilon)\sigma + \epsilon\tau). \tag{8.3}$$

This definition says that if a population of type σ is invaded by a small number of individuals of any other type τ, or if individuals of another type τ join the population because of a mutation, then individuals of type σ will have a better expected payoff against a random member of the mixed population than will individuals of type τ. Thus the invaders or mutants should die out.

Theorem 8.1. *A population state σ is evolutionarily stable if and only if for all $\tau \neq \sigma$,*

(1) $\pi_1(\sigma, \sigma) \geqslant \pi_1(\tau, \sigma)$.
(2) If $\pi_1(\tau, \sigma) = \pi_1(\sigma, \sigma)$, then $\pi_1(\sigma, \tau) > \pi_1(\tau, \tau)$.

The first condition says that (σ, σ) is a mixed strategy Nash equilibrium of the symmetric two-player game that is played in the evolutionary game. Therefore σ is a best response to σ. The second condition says that if τ is another best response to σ, then σ is a better response to τ than τ is to itself.

Proof. Assume σ is evolutionarily stable. Let $\tau \neq \sigma$. Then for all sufficiently small $\epsilon > 0$, (8.3) holds. Therefore, for all sufficiently small $\epsilon > 0$,

$$(1 - \epsilon)\pi_1(\sigma, \sigma) + \epsilon\pi_1(\sigma, \tau) > (1 - \epsilon)\pi_1(\tau, \sigma) + \epsilon\pi_1(\tau, \tau). \tag{8.4}$$

Letting $\epsilon \to 0$, we obtain condition (1). If $\pi_1(\tau, \sigma) = \pi_1(\sigma, \sigma)$, (8.4) becomes

$$\epsilon\pi_1(\sigma, \tau) > \epsilon\pi_1(\tau, \tau).$$

Dividing by ϵ, we obtain condition (2).

To prove the converse, assume that for all $\tau \neq \sigma$, conditions (1) and (2) are true. Consider a particular τ different from σ. Since conditions (1) holds, there are two possibilities.

(i) $\pi_1(\sigma, \sigma) > \pi_1(\tau, \sigma)$. Then it is easy to see that for small $\epsilon > 0$, (8.4) is true.
(ii) $\pi_1(\sigma, \sigma) = \pi_1(\tau, \sigma)$. Then (2) implies that $\pi_1(\sigma, \tau) > \pi_1(\tau, \tau)$. But then (8.4) holds for $0 < \epsilon < 1$.

Combining (i) and (ii), we see that σ is evolutionarily stable. $\qquad\square$

One consequence of Theorem 8.1 is the following theorem.

Theorem 8.2. *If (σ, σ) is a strict Nash equilibrium of a symmetric two-player game, then σ is an evolutionarily stable state of the corresponding evolutionary game.*

The reason is that for any Nash equilibrium, condition (1) holds; and for a strict Nash equilibrium, condition (2) is irrelevant. Of course, strict Nash equilibria use only pure strategies, so such populations consist entirely of individuals of one pure type i.

Another consequence of Theorem 8.1 is the following.

Theorem 8.3. *If σ is an evolutionarily stable state in which all pure strategies are active, then for all $\tau \neq \sigma$, (τ, τ) is not a Nash equilibrium of the symmetric two-player game. Hence there are no other evolutionarily stable states.*

Proof. According to Theorem 5.3, for such a σ we have that for all i, $\pi_1(s_i, \sigma) = \pi_1(\sigma, \sigma)$. Let $\tau = \sum q_i s_i$. Then

$$\pi_1(\tau, \sigma) = \sum_{i=1}^{n} q_i \pi_1(s_i, \sigma) = \sum_{1=1}^{n} q_i \pi_1(\sigma, \sigma) = \pi_1(\sigma, \sigma).$$

Therefore, since σ is evolutionarily stable, Theorem 8.1 (2) implies that for $\tau \neq \sigma$, $\pi_1(\sigma, \tau) > \pi_1(\tau, \tau)$. Therefore (τ, τ) is not a Nash equilibrium of the symmetric two-player game. $\qquad\square$

The same argument shows that the following result also holds.

Theorem 8.4. *Let $\sigma = \sum p_i s_i$ be an evolutionarily stable state, and let $I = \{i : p_i > 0\}$. Let $\tau \neq \sigma$, $\tau = \sum q_i s_i$, be a population state for which the the set of i values such that $q_i > 0$ is a subset of I. Then (τ, τ) is not a Nash equilibrium of the symmetric two-player game. Hence there are no other evolutionarily stable states in which the set of active strategies is a subset of the set of active strategies in σ.*

8.4 Evolutionary stability with two pure strategies

Consider an evolutionary game based on a symmetric two-player game in normal form with just two pure strategies. The payoff matrix must have the form

		Player 2	
		s_1	s_2
Player 1	s_1	(a, a)	(b, c)
	s_2	(c, b)	(d, d)

Theorem 8.5. *Suppose $a \neq c$ and $b \neq d$. There are four cases.*

(1) $a > c$ and $d < b$: strategy s_1 strictly dominates strategy s_2. There is one Nash equilibrium, (s_1, s_1). It is symmetric and strict, so the population state s_1 is evolutionarily stable.

(2) $a < c$ and $d > b$: strategy s_2 strictly dominates strategy s_1. There is one Nash equilibrium, (s_2, s_2). It is symmetric and strict, so the population state s_2 is evolutionarily stable.

To describe the other two cases, let

$$p = \frac{d - b}{(a - c) + (d - b)}, \quad so \quad 1 - p = \frac{a - c}{(a - c) + (d - b)}. \tag{8.5}$$

(3) $a > c$ and $d > b$: each strategy is the best response to itself. There are three Nash equilibria: (s_1, s_1), (s_2, s_2), and (σ, σ) with $\sigma = ps_1 + (1 - p)s_2$ and p given by (8.5). The first two are symmetric and strict, so the population states s_1 and s_2 are evolutionarily stable. The population state σ is not evolutionarily stable.

(4) $a < c$ and $d < b$: each strategy is the best response to the other strategy. There are three Nash equilibria: (s_1, s_2), (s_2, s_1), and (σ, σ) with $\sigma = ps_1 + (1 - p)s_2$ and p given by (8.5). Only the last is symmetric. The population state σ is evolutionarily stable.

Case (3) includes Stag Hunt. Case (4) includes Chicken.

Proof. You can find the pure strategy Nash equilibria by circling best responses.

To find mixed strategy Nash equilibria (σ, τ) with $\sigma = ps_1 + (1 - p)s_2$, $\tau = qs_1 + (1 - q)s_2$, we first add the probabilities to the payoff matrix:

			Player 2	
			q	$1 - q$
			s_1	s_2
Player 1	p	s_1	(a, a)	(b, c)
	$1 - p$	s_1	(c, b)	(d, d)

At least one player has two active strategies; suppose it is Player 2. Then if Player 2 uses either pure strategy s_1 or pure strategy s_2, she gets the same expected payoff when Player 1 uses σ. Therefore

$$pa + (1 - p)b = pc + (1 - p)d, \quad so \quad d - b = ((a - c) + (d - b))p.$$

Since $d - b \neq 0$ by assumption, we must have $(a - c) + (d - b) \neq 0$ to solve for p. Then p is given by (8.5). In cases (1) and (2), this value of p is not between (0) and (1), so it cannot be used. However, in cases (3) and (4) $0 < p < 1$, so both of Player 1's strategies are active. Then we can calculate q the same way. We find that $q = p$.

Now that we have a symmetric Nash equilibrium (σ, σ) in cases (3) and (4), we check whether the corresponding population state σ is evolutionarily stable. Since (σ, σ) is a Nash equilibrium, σ satisfies condition (1) of Theorem 8.1. Since both pure strategies are active in σ, by Theorem 5.3, every τ satisfies $\pi_1(\tau, \sigma) = \pi_1(\sigma, \sigma)$, so condition (2) must be checked for every $\tau \neq \sigma$. For $\tau = qs_1 + (1 - q)s_2$, we calculate

$$
\begin{aligned}
\pi_1(\sigma, \tau) - \pi_1(\tau, \tau) &= paq + pb(1 - q) + (1 - p)cq + (1 - p)d(1 - q) \\
&\quad - (qaq + qb(1 - q) + (1 - q)cq + (1 - q)d(1 - q)) \\
&= (p - q)(aq + b(1 - q) - cq - d(1 - q)) \\
&= (p - q)(b - d + ((a - c) + (d - b))q) \\
&= (p - q)((a - c) + (d - b))\left(\frac{b - d}{(a - c) + (d - b)} + q\right) \\
&= -(p - q)((a - c) + (d - b))\left(\frac{d - b}{(a - c) + (d - b)} - q\right) \\
&= -(p - q)^2((a - c) + (d - b)).
\end{aligned}
$$

If $\tau \neq \sigma$, then $q \neq p$, so $(p - q)^2 > 0$. Thus we see that in case (3) (in which $a - c$ and $d - b$ are both positive), $\pi_1(\sigma, \tau) - \pi_1(\tau, \tau) < 0$ for all $\tau \neq \sigma$, so σ is *not* evolutionarily stable; and in case (4) (in which $a - c$ and $d - b$ are both negative), $\pi_1(\sigma, \tau) - \pi_1(\tau, \tau) > 0$ for all $\tau \neq \sigma$, so σ *is* evolutionarily stable. $\qquad\square$

In case (3), the population state σ is the opposite of evolutionarily stable: if τ is any invading population type, σ does worse against τ than τ does against itself.

8.4.1 Stag Hunt.

Consider an evolutionary game based on Stag Hunt (Subsection 3.2.2). The payoff matrix is reproduced below.

		Hunter 2	
		stag	hare
Hunter 1	stag	$(2, 2)$	$(0, 1)$
	hare	$(1, 0)$	$(1, 1)$

In Theorem 8.5, we are in case (3). There are two symmetric pure-strategy strict Nash equilibria, (stag, stag) and (hare, hare). Both pure populations, all stag hunters and all hare hunters, are evolutionarily stable. There is also a symmetric mixed-strategy Nash equilibrium in which each player uses the strategy stag half the time and the strategy hare half the time. However, the corresponding population state is not evolutionarily stable.

8.4.2 Stag Hunt variation. Suppose in the game of Stag Hunt, a hunter who hunts the stag without help from the other hunter has a $\frac{1}{4}$ chance of catching it. (Previously we assumed he had no chance of catching it.) Then the payoff matrix becomes

		Hunter 2	
		stag	hare
Hunter 1	stag	$(2,2)$	$(1,1)$
	hare	$(1,1)$	$(1,1)$

The corresponding evolutionary game is not covered by Theorem 8.5, because $d = b = 1$. There are two symmetric pure-strategy Nash equilibria, (stag, stag) and (hare, hare). However, only (stag, stag) is a strict Nash equilibrium. Indeed, the strategy hare is now weakly dominated by the strategy stag. There are no mixed strategy Nash equilibria.

By Theorem 8.2, the pure population consisting of all stag hunters is evolutionarily stable.

What about the pure population consisting of all hare hunters? Since $\pi_1(s,h) = \pi_1(h,h) = 1$, when we check condition (2) of Theorem 8.1, among the strategies τ that must be checked is the pure strategy, hunt stags. However, $\pi_1(h,s) = 1$ and $\pi_1(s,s) = 2$; that is, if some stag hunters invade the population of hare hunters, they do better against themselves than the hare hunters do against them. Thus a pure population of hare hunters is not evolutionarily stable.

8.4.3 Hawks and Doves. Consider a population of animals that fights over food, territory, or mates. We consider two possible strategies:

- Hawk (h): fight until either you are injured or your opponent retreats.
- Dove (d): display hostility, but if your opponent won't retreat, you retreat.

Let

- v = value of what you are fighting over.
- w = cost of injury.
- t = cost of protracted display.

We assume v, w, and t are all positive, and $v < w$. The payoff matrix is

		Animal 2	
		h	d
Animal 1	h	$(\frac{v-w}{2}, \frac{v-w}{2})$	$(v,0)$
	d	$(0,v)$	$(\frac{v}{2}-t, \frac{v}{2}-t)$

In Theorem 8.5, we are in case (4). Thus there are no symmetric pure-strategy Nash equilibria, and there is a mixed-strategy Nash equilibrium (σ, σ), $\sigma = ph + (1 - p)d$; you can check that $p = (v + 2t)/(w + 2t)$. The population state σ is evolutionarily stable.

The payoff to Animal 1 at the Nash equilibrium is can be computed by calculating $\pi_1(d, \sigma)$:

$$\pi_1(d, \sigma) = p \cdot 0 + (1 - p)\left(\frac{v}{2} - t\right) = \left(1 - \frac{v + 2t}{w + 2t}\right)\left(\frac{v}{2} - t\right).$$

From this expression we see that for fixed v and t, as w (the cost of injury) increases, the expected payoff increases. The reason is that as w increases, p (the probability that the Hawk strategy is used) decreases. When w is only a little bigger than v, p is close to 1, so the Hawk strategy is used a lot, and injuries frequently occur. Injuries occur less often at the Nash equilibrium when the expected result of fighting is very costly injuries.

For other ways that have evolved to minimize fighting among animals, see Problem 8.6.4 and Section 10.11.

For more information about Hawks and Doves, see the Wikipedia page http://en.wikipedia.org/wiki/Hawk-dove_game.

8.5   Sex Ratio 2

Recall the sex ratio pseudogame analyzed in Subsection 7.3.1. There was no actual game. There was, however, a situation very close to that considered in this chapter. The female population as a whole produces a fraction v of male offspring and a fraction $1 - v$ of female offspring. The number v can be regarded as the population state. An individual female produces a fraction u of male offspring and a fraction $1 - u$ of female offspring. The number u can be regarded as the type of an individual. The payoff to this individual $\pi_1(u, v)$ is her number of grandchildren. We derived the formula

$$\pi_1(u, v) = \sigma_f c^2 \left(1 + u\frac{1 - 2v}{v}\right).$$

In this situation, our analog of a Nash equilibrium was the pair $(\frac{1}{2}, \frac{1}{2})$, in the sense that if the population state was $\frac{1}{2}$ (i.e., females as a whole have $\frac{1}{2}$ male offspring), an individual could do no better than by choosing also to have $\frac{1}{2}$ male offspring.

Is this population state evolutionarily stable? We saw that for any individual type u, we have

$$\pi_1(u, \tfrac{1}{2}) = \pi_1(\tfrac{1}{2}, \tfrac{1}{2}) = \sigma_f c^2.$$

Thus we must check condition (2) of Theorem 8.1 for every $u \neq \frac{1}{2}$. We have

$$\pi_1(\tfrac{1}{2}, u) - \pi_1(u, u) = \sigma_f c^2 \left(1 + \frac{1}{2}\frac{1 - 2u}{u}\right) - \sigma_f c^2 \left(1 + u\frac{1 - 2u}{u}\right)$$

$$= \sigma_f c^2 (\tfrac{1}{2} - u)\frac{1 - 2u}{u} = \frac{2\sigma_f c^2}{u}(\tfrac{1}{2} - u)^2.$$

Since this is positive for $u \neq \frac{1}{2}$, the population state $\frac{1}{2}$ is evolutionarily stable.

8.6 Problems

8.6.1 Two Fishermen.
In a certain fishing village, two fishermen own nets that are put out in the evening. The two share the catch equally regardless of whether they help put out the nets. The value of the expected catch is v, and the cost to each fisherman of putting out the nets if they do it together is c_2. However, if one fisherman puts out the nets by himself, his cost is c_1, and the cost to the other fisherman is 0. Assume that $\frac{v}{2} > c_1 > c_2$. The normal form of this game is shown in the following payoff matrix:

		Fisherman 2	
		help	**don't help**
Fisherman 1	**help**	$(\frac{v}{2} - c_2, \frac{v}{2} - c_2)$	$(\frac{v}{2} - c_1, \frac{v}{2})$
	don't help	$(\frac{v}{2}, \frac{v}{2} - c_1)$	$(0, 0)$

(1) Use best response to find pure-strategy Nash equilibria.
(2) Use Theorem 8.5 to find the mixed strategy Nash equilibrium and the evolutionarily stable state of the corresponding evolutionary game.

8.6.2 Lions and Impalas 1.
Two lions see a big impala and a little impala in the distance. Each lion independently chooses which impala to chase. The lions will kill whichever impala they chase, but if they choose the same impala, they will have to share it. The value of the big impala is 4, the value of the little impala is 2. The payoff matrix is then

		lion 2	
		big	**little**
lion 1	**big**	(2, 2)	(4, 2)
	little	(2, 4)	(1, 1)

Theorem 8.5 does not apply, because $a = c$.

(1) Are there any strictly dominated or weakly dominated strategies?
(2) Find the pure strategy Nash equilibria.
(3) Check whether any pure strategy symmetric Nash equilibria that you found in part (2) correspond to evolutionarily stable states.

8.6.3 Stag Hunt with Easily Spooked Hares 1. Two hunters working together can kill a stag, which is worth 8. One hunter alone cannot kill a stag. One hunter alone can kill four hares, which are each worth 1. However, if two hunters go after the hares, the hares will run away, and each hunter will only kill one hare.

If two hunters are hunting together, each has two possible strategies, hunt a stag (s) or or hunt the hares (h). The payoffs are as follows:

		Hunter 2	
		s	h
Hunter 1	s	$(4,4)$	$(0,4)$
	h	$(4,0)$	$(1,1)$

Theorem 8.5 does not apply, because $a = c$.

(1) Use best response to find the pure strategy Nash equilibria.
(2) Which of the pure strategy Nash equilibria correspond to an evolutionarily stable state?

8.6.4 Hawks, Doves, and Protecters. In the game of Hawks and Doves (Section 8.4.3), let's add a new strategy, Protecters: when you meet another animal, if he was there first, use the Dove strategy, but if you were there first, use the Hawk strategy.

In this problem, assume v, w, and t are all positive.

(1) Explain why the following payoff matrix is plausible:

		Animal 2		
		h	d	p
Animal 1	h	$(\frac{v-w}{2}, \frac{v-w}{2})$	$(v,0)$	$(\frac{3v-w}{4}, \frac{v-w}{4})$
	d	$(0,v)$	$(\frac{v}{2} - t, \frac{v}{2} - t)$	$(\frac{v}{4} - \frac{t}{2}, \frac{3v}{4} - \frac{t}{2})$
	p	$(\frac{v-w}{4}, \frac{3v-w}{4})$	$(\frac{3v}{4} - \frac{t}{2}, \frac{v}{4} - \frac{t}{2})$	$(\frac{v}{2}, \frac{v}{2})$

(2) Show that if $v < w$, then p is an evolutionarily stable state of the corresponding evolutionary game. (This state is in fact often observed in nature. For example, birds typically build their own nests and are willing to defend them. However, no fighting occurs, because birds do not try to take other birds' nests.)
(3) Show that if $v = w$, then (p,p) is a Nash equilibrium, but p is not an evolutionarily stable state.

8.6.5   A correlated equilibrium in the game of Hawks and Doves.   Consider the game of Hawks and Doves in Section 8.4.3 with $v < w$. When two animals meet, they both observe which of them arrived first. This is a public signal. The animal that arrives first uses h, and the other animal uses d. Explain why this is a correlated equilibrium. Assuming each animal arrives first half the time, show that the expected payoff to each animal in the correlated equilibrium is greater than the expected payoff to each animal in the mixed-strategy Nash equilibrium. (This is a traditional explanation of why societies have property rights.)

Chapter 9

Differential equations

In Chapter 10 we investigate how strategies in evolutionary games change over time, with more successful strategies displacing less successful ones. Our study will use *differential equations*, the branch of mathematics that deals with quantities that change continuously in time. This chapter introduces the point of view and tools of differential equations that will be needed.

9.1 Differential equations and scientific laws

Suppose $x(t) = (x_1(t), \ldots, x_n(t))$ is a moving point in $\mathbb{R}^n$. At time t, its velocity vector is $\dot{x}(t) = (\dot{x}_1(t), \ldots, \dot{x}_n(t))$. (Note that we use a dot to indicate derivative with respect to t.) The velocity vector is usually drawn with its tail at the point $x(t)$.

For example, suppose $x(t) = (\cos t, \sin t)$, a moving point in $\mathbb{R}^2$. The point $x(t)$ runs around the circle of radius 1, centered at the origin. We have $\dot{x}(t) = (-\sin t, \cos t)$. Therefore $x(0) = (1,0)$, $\dot{x}(0) = (0,1)$, $x(\frac{\pi}{2}) = (0,1)$, and $\dot{x}(\frac{\pi}{2}) = (-1,0)$. These facts are illustrated in Figure 9.1.

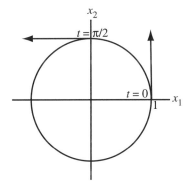

Figure 9.1.

Often a scientific law tells us: if you know x, a point that represents the state of the system, at some time, then you know $\dot{x}$, how x is changing, at that time. In other words, the velocity vector $\dot{x}$ is a function of the state x, that is, $\dot{x} = f(x)$ or

$$\dot{x}_1 = f_1(x_1, \ldots, x_n), \tag{9.1}$$

$$\vdots$$

$$\dot{x}_n = f_n(x_1, \ldots, x_n). \tag{9.2}$$

An equation of the form $\dot{x} = f(x)$ is a *first order autonomous ordinary differential equation*:

- first order: there are only first derivatives, not higher derivatives;
- autonomous: the derivative only depends on the state of the system x, not on the time t; and
- ordinary: there are only ordinary derivatives, not partial derivatives.

When a differential equation $\dot{x} = f(x)$ on $\mathbb{R}^n$ with $n > 1$ is written in the form (9.1)–(9.2), it is sometimes called a *system of differential equations*.

To use the scientific law or differential equation to make a prediction of what will happen (i.e., to predict $x(t)$), we need to solve an *initial value problem*:

$$\dot{x} = f(x), \quad x(t_0) = x_0.$$

In other words, given the differential equation $\dot{x} = f(x)$ and the state of the system at time t_0, $x(t_0) = x_0$, we need to find a function $x(t)$ such that $x(t_0) = x_0$ and, at every time t, $\dot{x}(t) = f(x(t))$.

For example, the system

$$\dot{x}_1 = -x_2, \tag{9.3}$$

$$\dot{x}_2 = x_1, \tag{9.4}$$

with the initial condition $(x_1(0), x_2(0)) = (1, 0)$, has the solution $(x_1(t), x_2(t)) = (\cos t, \sin t)$. To check that this is indeed a solution of the system, just substitute $\dot{x}_1(t)$ and $\dot{x}_2(t)$ into the left side, and substitute $x_1(t)$ and $x_2(t)$ into the right side:

$$-\sin t = -\sin t,$$

$$\cos t = \cos t.$$

To check that $(x_1(0), x_2(0)) = (1, 0)$, just notice that $\cos 0 = 1$ and $\sin 0 = 0$.

The following theorem gathers some fundamental facts about differential equations.

Theorem 9.1. *Let U be an open set in $\mathbb{R}^n$, let $f : U \to \mathbb{R}^n$ be a continuously differentiable function, and let $x_0 \in U$. Then:*

(1) The initial value problem

$$\dot{x} = f(x), \quad x(t_0) = x_0,$$

has a unique solution.

(2) If $x(t)$ stays bounded and stays away from the boundary of U as t increases (respectively decreases), then $x(t)$ is defined for $t_0 \leqslant t < \infty$ (respectively $-\infty < t \leqslant t_0$).

When we consider differential equations $\dot{x} = f(x)$, we always assume that f is continuously differentiable, so that this theorem applies.

The set U on which the differential equation is defined is called *phase space*.

A point x_0 at which $f(x_0) = 0$ is an *equilibrium* of $\dot{x} = f(x)$. If x_0 is an equilibrium of $\dot{x} = f(x)$, then the unique solution of the initial value problem

$$\dot{x} = f(x), \quad x(t_0) = x_0,$$

is $x(t) = x_0$ for $-\infty < t < \infty$. To prove this, just check that the formula for $x(t)$ gives a solution of the initial value problem, and recall that solutions are unique.

9.2   The phase line

Example. A population increases with *growth rate* 5% per year. The rate of change of the population is the growth rate times the population. Therefore, if x = population and t = time in years, then $\dot{x} = .05x$. The solution with $x(0) = x_0$ is $x = x_0 e^{.05t}$. Notice that

- All solutions approach 0 as $t \to -\infty$.
- The solution with $x(0) = 0$ is $x(t) = 0$ for $-\infty < t < \infty$.

In general, the solution of $\dot{x} = rx$, r a constant, with $x(0) = x_0$, is $x = x_0 e^{rt}$. If $r > 0$, all solutions approach 0 as $t \to -\infty$. If $r < 0$, all solutions approach 0 as $t \to \infty$.

One way to see this geometrically is by drawing the *phase line*, which is the x-axis with dots where equilibria are located and arrows to show where solutions are increasing and decreasing. See Figure 9.2. Where $\dot{x} > 0$, $x(t)$ is increasing; where $\dot{x} < 0$, $x(t)$ is decreasing (shown by the arrows).

Example. A population x has growth rate $r(1 - \frac{x}{c})$; r and c are positive constants. (This time growth rate is expressed as a number rather than as a percentage.) Notice that the growth rate is positive for $0 < x < c$ and negative for $x > c$. The number c is the *carrying capacity* of the environment. The

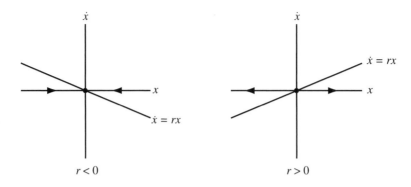

Figure 9.2. Phase lines for $\dot{x} = rx$. The graph of $\dot{x} = rx$
helps when drawing the phase line.

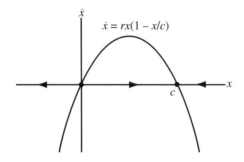

Figure 9.3. Phase line for $\dot{x} = rx\left(1 - \frac{x}{c}\right)$. The graph of $\dot{x} = rx\left(1 - \frac{x}{c}\right)$
helps when drawing the phase line.

rate of change of the population is the growth rate times the population, so $\dot{x} = rx\left(1 - \frac{x}{c}\right)$. The phase line is shown in Figure 9.3.

The region $x > 0$ is called the *basin of attraction* of the equilibrium $x = c$: if a solution starts in that region, it approaches the equilibrium $x = c$ as $t \to \infty$.

9.3 Vector fields

Geometrically, the differential equation $\dot{x} = f(x)$, with $x \in \mathbb{R}^n$ and f a function from an open set U in $\mathbb{R}^n$ to $\mathbb{R}^n$, defines a *vector field* on U. The vector $f(x)$ at the point x is drawn with its tail at x.

If you sketch a few vectors for the system (9.3)–(9.4), you will quickly see that solutions should wind around the origin in the counterclockwise direction. The solution $(x_1(t), x_2(t)) = (\cos t, \sin t)$ that we found does exactly that. We next give a more complicated example.

Example. Let x be the population of a prey species and y the population of a predator species. We assume that the growth rate of x is $a - by$ and the growth rate of y is $-c + ex$, where $a, b, c,$ and e are positive constants. These formulas are reasonable: the growth rate of the prey is positive when there are no predators, and is lower when there are more predators; the growth rate of the predators is negative when there are no prey, and is higher when there are more prey. The resulting system is

$$\dot{x} = x(a - by), \tag{9.5}$$
$$\dot{y} = y(-c + ex). \tag{9.6}$$

To sketch the vector field, we first find the *nullclines*, which are the curves where $\dot{x} = 0$ or $\dot{y} = 0$:

- $\dot{x} = 0$ if $x = 0$ or $y = \frac{a}{b}$ (two lines).
- $\dot{y} = 0$ if $y = 0$ or $x = \frac{c}{e}$ (two lines).

Where an $\dot{x} = 0$ nullcline meets a $\dot{y} = 0$ nullcline, there is an equilibrium. The equilibria divide the nullclines into open curves. On each such curve:

- For an $\dot{x} = 0$ nullcline, the vectors are vertical. Check a point to see whether the vectors point up or down.
- For a $\dot{y} = 0$ nullcline, the vectors are horizontal. Check a point to see whether the vectors point right or left.

The nullclines divide the plane into open regions on which $\dot{x}$ and $\dot{y}$ do not change sign. The signs of $\dot{x}$ and $\dot{y}$ in a region can be determined by checking one point in the region. These signs determine whether the vectors in that region point northeast, northwest, southwest, or southeast.

Figure 9.4. Nullclines, equilibria, and vector field for (9.5)–(9.6).

The vector field for the system (9.5)–(9.6) is shown in Figure 9.4. Only the first quadrant ($x \geqslant 0$, $y \geqslant 0$) is physically relevant. The populations are in equilibrium at $(x, y) = (0, 0)$ (not very interesting) and $(x, y) = \left(\frac{c}{e}, \frac{a}{b}\right)$. In the open first quadrant ($x > 0$, $y > 0$), solutions appear to go around the equilibrium $(\frac{c}{e}, \frac{a}{b})$ in the counterclockwise direction, but it is impossible to tell from the picture whether they spiral in, spiral out, or rejoin themselves to form time-periodic solutions.

9.4 Functions and differential equations

Consider the differential equation $\dot{x} = f(x)$ on $\mathbb{R}^n$. Let $x(t)$ be a solution. Let $V: \mathbb{R}^n \to \mathbb{R}$ be a continuously differentiable function. Then $V(x(t))$ gives the value of V along the solution as a function of t. According to the chain rule, the rate of change of V is

$$\dot{V} = \frac{\partial V}{\partial x_1}(x(t))\,\dot{x}_1(t) + \cdots + \frac{\partial V}{\partial x_n}(x(t))\,\dot{x}_n(t) = \nabla V(x(t)) \cdot \dot{x}(t),$$

where

$$\nabla V(x) = \left(\frac{\partial V}{\partial x_1}(x), \ldots, \frac{\partial V}{\partial x_n}(x)\right)$$

is the gradient of V at the point x, and $\cdot$ represents dot product.

This formula has many uses.

9.4.1 Invariant curves and surfaces. Suppose that $V = c$ implies $\dot{V} = 0$. Then any solution $x(t)$ with $V = c$ at some time t_0 has $V(x(t)) = c$ for all time. (This conclusion is true provided $\nabla V \neq 0$ on the set $V = c$; this technical requirement normally holds, and we will ignore it.) In other words, the set $V = c$ is *invariant*: if you start on it, you stay on it. In two dimensions, a set of the form $V(x_1, x_2) = c$ is a curve. In three dimensions, a set of the form $V(x_1, x_2, x_3) = c$ is a surface.

For example, in the predator-prey system (9.5)–(9.6), note that if $x = 0$, then $\dot{x} = 0$. This implies that the line $x = 0$ (the y-axis) is invariant. In other words, if the prey population starts at 0, it stays at 0. Note that in this case the predator population decays to 0, since there is nothing for the predators to eat.

In the predator-prey system note also that if $y = 0$, then $\dot{y} = 0$, so the line $y = 0$ is invariant. In other words, if the predator population starts at 0, it stays at 0. In this case the prey population increases without bound.

9.4.2 First integrals. Suppose $\dot{V} = 0$ everywhere. Then all sets $V = $ constant are invariant. In this case V is called a *first integral* of the differential equation.

For a differential equation on $\mathbb{R}^2$ such as the predator-prey system (9.5)–(9.6), one can try to find a first integral by the following procedure: Divide the equation for $\dot{y}$ by the equation for $\dot{x}$, yielding an equation of the form $dy/dx = g(x,y)$. Try to solve this differential equation to obtain a general solution of the form $V(x,y) = c$. Then the function V is a first integral.

Let's try this for (9.5)-(9.6). Dividing (9.6) by (9.5), we obtain

$$\frac{dy}{dx} = \frac{y(-c+ex)}{x(a-by)}.$$

This system can be solved by separation of variables. Some algebra yields

$$\frac{a-by}{y}\,dy = \frac{-c+ex}{x}\,dx \quad \text{or} \quad \left(\frac{a}{y}-b\right)dy = \left(-\frac{c}{x}+e\right)dx.$$

We integrate both sides and add an arbitrary constant k to one side. In the open first quadrant, where x and y are positive, we obtain

$$a\ln y - by + k = -c\ln x + ex \quad \text{or} \quad ex - c\ln x + by - a\ln y = k.$$

Hence the function $V(x,y) = ex - c\ln x + by - a\ln y$ is a first integral for (9.5)-(9.6) in the open first quadrant.

Note that $V_x = e - \frac{c}{x}$ and $V_y = b - \frac{a}{y}$. The function V has a critical point where $V_x = 0$ and $V_y = 0$. This occurs at $(x,y) = (\frac{c}{e}, \frac{a}{b})$, which, we recall, is an equilibrium of (9.5)-(9.6). Actually, V has a local minimum at that point. We can check this by checking that $V_{xx}V_{yy} - V_{xy}^2 > 0$ and $V_{xx} > 0$ at that point:

$$V_{xx} = \frac{c}{x^2}, \quad V_{yy} = \frac{a}{y^2}, \quad V_{xy} = 0 \quad \text{so} \quad V_{xx}V_{yy} - V_{xy}^2 = \frac{ac}{x^2 y^2}.$$

Therefore at any point in the first quadrant (in particular the point in question), $V_{xx}V_{yy} - V_{xy}^2 > 0$ and $V_{xx} > 0$.

Since V has a local minimum at $(\frac{c}{e}, \frac{a}{b})$, level curves of V near that point are closed. (The same conclusion would hold if V had a local maximum there, but it would not hold at a saddle point.) The level curves of V are invariant under (9.5)-(9.6). Thus in Figure 9.4, near the equilibrium $(\frac{c}{e}, \frac{a}{b})$, solutions go around the equilibrium counterclockwise, ending exactly where they started, then go around again: they are time periodic.

Actually, all solutions in the open first quadrant rejoin themselves, but we will not show this.

An *orbit* of a differential equation is the curve in phase space that is traced out by a solution. A *phase portrait* of $\dot{x} = f(x)$ is a sketch of phase space that shows all unusual orbits and examples of typical orbits, together with arrows on the orbits that indicate the direction of movement.

The phase lines of Figures 9.2 and 9.3 are phase portraits. The phase line of Figure 9.3 has exactly five orbits: $(-\infty, 0)$, $\{0\}$, $(0,c)$, $\{c\}$, and (c,∞). In one dimension the number of orbits is typically finite, so all can be shown.

The phase portrait of the predator-prey system (9.5)–(9.6) is shown in Figure 9.5. When the dimension is greater than one, the number of orbits is infinite, so not all can be shown. The unusual orbits are the equilibria and orbits on the x- and y-axes; these are all shown. (In this phase portrait, the orbits on the x- and y-axes are unusual in that they approach an equilibrium in one direction.) In each quadrant, only examples of typical orbits are shown. We have seen that in the open first quadrant, all orbits surround the equilibrium. Just two are shown.

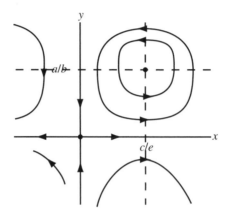

Figure 9.5. Phase portrait of the predator-prey system (9.5)–(9.6). Nullclines that are not orbits are dashed. Compare Figure 9.4.

9.4.3 Stability and Lyapunov functions.

An equilibrium x_0 of $\dot{x} = f(x)$ is *stable* if solutions that start near x_0 stay near x_0 in future time. More precisely, x_0 is stable if for each $\epsilon > 0$ there exists $\delta > 0$ such that if $\|x_1 - x_0\| < \delta$, then the solution $x(t)$ with $x(0) = x_1$ satisfies

$$\|x(t) - x_0\| < \epsilon \quad \text{for all } t \geqslant 0. \tag{9.7}$$

An equilibrium x_0 of $\dot{x} = f(x)$ is *asymptotically stable* if solutions that start near x_0 stay near x_0 in future time, and in addition approach x_0 as $t \to \infty$. More precisely, x_0 is asymptotically stable if for each $\epsilon > 0$ there exists $\delta > 0$ such that if $\|x_1 - x_0\| < \delta$, then (i) the solution $x(t)$ with $x(0) = x_1$ satisfies (9.7), and (ii) $x(t)$ approaches x_0 as $t \to \infty$.

Asymptotically stable equilibria are states that one expects to observe persisting in the natural world. If some perturbation takes the state of the system a small distance away from an asymptotically stable equilibrium, the state returns to the equilibrium.

In Figure 9.3, the equilibrium c is asymptotically stable (and hence stable). The equilibrium 0 is not stable (and hence not asymptotically stable). In Figure 9.5, the equilibrium $(\frac{c}{e}, \frac{a}{b})$ is stable but not asymptotically stable. The equilibrium $(0, 0)$ is not stable (and hence not asymptotically stable).

Theorem 9.2. *Lyapunov's Theorem. Let x_0 be an equilibrium of $\dot{x} = f(x)$, U an open set in $\mathbb{R}^n$ that contains x_0, and $V : U \to \mathbb{R}$ a continuously differentiable function such that $V(x_0) = 0$ and $V(x) > 0$ for $x \neq x_0$.*

(1) If $\dot{V} \leqslant 0$ for all x, then x_0 is stable.

(2) If $\dot{V} < 0$ for all $x \neq x_0$, then x_0 is asymptotically stable.

The function V is called a *Lyapunov function* in the first case, and a *strict Lyapunov function* in the second case.

We remark that at the point x_0, $\dot{V}$ is always 0.

Example. For constants a and r, consider the system

$$\dot{x} = -rx - ay,$$
$$\dot{y} = ax - ry.$$

There is an equilibrium at the origin. Let $V(x, y) = x^2 + y^2$. The function $V(x, y)$ satisfies $V(0, 0) = 0$ and $V(x, y) > 0$ for $(x, y) \neq (0, 0)$. We calculate

$$\dot{V} = \frac{\partial V}{\partial x}\dot{x} + \frac{\partial V}{\partial y}\dot{y} = 2x(-rx - ay) + 2y(ax - ry) = -2r(x^2 + y^2).$$

By Theorem 9.2, the origin is stable if $r = 0$ and is asymptotically stable if $r > 0$. In the case $r = 0$, V is actually a first integral.

9.5 Linear differential equations

A *linear differential equation* is a system of the form

$$\dot{x}_1 = a_{11}x_1 + a_{12}x_2 + \cdots + a_{1n}x_n, \tag{9.8}$$

$$\dot{x}_2 = a_{21}x_1 + a_{22}x_2 + \cdots + a_{2n}x_n, \tag{9.9}$$

$$\vdots$$

$$\dot{x}_n = a_{n1}x_1 + a_{n2}x_2 + \cdots + a_{nn}x_n, \tag{9.10}$$

with all the a_{ij}s constants.

Let

$$x = \begin{pmatrix} x_1 \\ x_2 \\ \vdots \\ x_n \end{pmatrix}, \quad \dot{x} = \begin{pmatrix} \dot{x}_1 \\ \dot{x}_2 \\ \vdots \\ \dot{x}_n \end{pmatrix}, \quad A = \begin{pmatrix} a_{11} & a_{12} & \cdots & a_{1n} \\ a_{21} & a_{22} & \cdots & a_{2n} \\ \vdots & \vdots & \ddots & \vdots \\ a_{n1} & a_{n2} & \cdots & a_{nn} \end{pmatrix}.$$

Then the system (9.8)–(9.10) can be written as the single *matrix differential equation* $\dot{x} = Ax$ (matrix product).

In the case $n = 1$, (9.8)–(9.10) reduces to $\dot{x} = ax$, with $x \in \mathbb{R}$ and a a constant. The solution with $x(0) = x_0$ is $x = x_0 e^{at}$.

With this example in mind, it is reasonable to ask whether the matrix differential equation $\dot{x} = Ax$ has any solutions of the form $x = x_0 e^{\lambda t}$. (Here x and x_0 are in $\mathbb{R}^n$, λ is a constant, and x_0 should be a nonzero vector to get an interesting result.) To answer this question, we substitute $x = x_0 e^{\lambda t}$ into both sides of $\dot{x} = Ax$ and obtain

$$\lambda e^{\lambda t} x_0 = A e^{\lambda t} x_0 \quad \text{or} \quad \lambda x_0 = A x_0 \quad \text{or} \quad (A - \lambda I) x_0 = 0.$$

Here I is the $n \times n$ identity matrix.

$$I = \begin{pmatrix} 1 & 0 & 0 & \cdots & 0 \\ 0 & 1 & 0 & \cdots & 0 \\ \vdots & \vdots & \vdots & \ddots & \vdots \\ 0 & 0 & 0 & \cdots & 1 \end{pmatrix},$$

which has the property $Ix = x$ for any $x \in \mathbb{R}^n$.

The equation $(A - \lambda I) x_0 = 0$ has solutions other than $x_0 = 0$ if and only if the determinant of the matrix $A - \lambda I$ is 0, that is, $\det(A - \lambda I) = 0$. The numbers λ such that $\det(A - \lambda I) = 0$ are called *eigenvalues* of A. Corresponding vectors x_0 such that $(A - \lambda I) x_0 = 0$ are called *eigenvectors*. Eigenvalues and eigenvectors may be complex. The equation $\det(A - \lambda I) = 0$ turns out to be a polynomial equation of degree n in λ (the *characteristic equation* of A), so there are exactly n eigenvalues, counting multiplicity.

Example. Consider the linear system

$$\dot{x} = y, \tag{9.11}$$

$$\dot{y} = x. \tag{9.12}$$

Written as a matrix equation, it is

$$\begin{pmatrix} \dot{x} \\ \dot{y} \end{pmatrix} = \begin{pmatrix} 0 & 1 \\ 1 & 0 \end{pmatrix} \begin{pmatrix} x \\ y \end{pmatrix}.$$

The characteristic equation is

$$\det\left(\begin{pmatrix} 0 & 1 \\ 1 & 0 \end{pmatrix} - \lambda \begin{pmatrix} 1 & 0 \\ 0 & 1 \end{pmatrix} \right) = \det \begin{pmatrix} -\lambda & 1 \\ 1 & -\lambda \end{pmatrix} = \lambda^2 - 1 = 0.$$

Therefore the eigenvalues are $\lambda = \pm 1$.

To find eigenvectors for the eigenvalue $\lambda = -1$, we look for solutions to the equation $(A - (-1)I)x_0 = 0$, with $A = \left(\begin{smallmatrix} 0 & 1 \\ 1 & 0 \end{smallmatrix}\right)$:

$$\left(\begin{pmatrix} 0 & 1 \\ 1 & 0 \end{pmatrix} - (-1)\begin{pmatrix} 1 & 0 \\ 0 & 1 \end{pmatrix}\right)\begin{pmatrix} x \\ y \end{pmatrix} = \begin{pmatrix} 0 \\ 0 \end{pmatrix} \quad \text{or} \quad \begin{pmatrix} 1 & 1 \\ 1 & 1 \end{pmatrix}\begin{pmatrix} x \\ y \end{pmatrix} = \begin{pmatrix} 0 \\ 0 \end{pmatrix}.$$

The solutions of this equation are all multiples of the vector $\left(\begin{smallmatrix} -1 \\ 1 \end{smallmatrix}\right)$. These are the eigenvectors for the eigenvalue -1. If $x(0) = x_0$ is a nonzero multiple of $\left(\begin{smallmatrix} 1 \\ -1 \end{smallmatrix}\right)$, then $x(t) = e^{-t}x_0$, so $x(t)$ is a positive multiple of x_0 for all t, $x(t) \to 0$ as $t \to \infty$, and $\|x(t)\| \to \infty$ as $t \to -\infty$.

Similarly, for the eigenvalue $\lambda = 1$, the eigenvectors are all multiples of the vector $\left(\begin{smallmatrix} 1 \\ 1 \end{smallmatrix}\right)$. If $x(0) = x_0$ is a nonzero multiple of $\left(\begin{smallmatrix} 1 \\ 1 \end{smallmatrix}\right)$, then $x(t) = e^{t}x_0$, so $x(t)$ is a positive multiple of x_0 for all t, $x(t) \to 0$ as $t \to -\infty$, and $\|x(t)\| \to \infty$ as $t \to \infty$.

Using this information, the phase portrait of the linear system (9.11)-(9.12) can be sketched; see Figure 9.6. The line $y = -x$ consists of eigenvectors for the eigenvalue -1; on it the direction of movement is toward the origin. The line $y = x$ consists of eigenvectors for the eigenvalue 1; on it the direction of movement is away from the origin. Other initial conditions can be regarded as a linear combination of $\left(\begin{smallmatrix} -1 \\ 1 \end{smallmatrix}\right)$ and $\left(\begin{smallmatrix} 1 \\ 1 \end{smallmatrix}\right)$. As t increases, the component in the $\left(\begin{smallmatrix} -1 \\ 1 \end{smallmatrix}\right)$ direction decreases, while the component in the $\left(\begin{smallmatrix} 1 \\ 1 \end{smallmatrix}\right)$ direction increases.

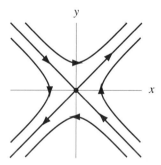

Figure 9.6. Phase portrait of the linear system (9.11)-(9.12).

Of course, the phase portrait of (9.11)-(9.12) could also be sketched by first sketching nullclines. Alternatively, it could be sketched by finding a first integral, for example, $x^2 - y^2$.

The linear differential equation $\dot{x} = Ax$ is called *hyperbolic* if all eigenvalues of A have nonzero real part. There are three cases:

- All eigenvalues have negative real part: all solutions approach the origin as $t \to \infty$. The origin is asymptotically stable.

- All eigenvalues have positive real part: all solutions approach the origin as $t \to -\infty$. The origin is asymptotically stable for $\dot{x} = -Ax$.
- Counting multiplicity, k eigenvalues have negative real part and $n - k$ eigenvalues have positive real part. Then there are subspaces E^s of dimension k and E^u of dimension $n - k$ such that:

  — a solution $x(t)$ of $\dot{x} = Ax$ approaches the origin as $t \to \infty$ if and only if $x(0) \in E^s$;

  — a solution $x(t)$ of $\dot{x} = Ax$ approaches the origin as $t \to -\infty$ if and only if $x(0) \in E^u$.

E^s and E^u are called the *stable subspace* and the *unstable subspace*, respectively, of $\dot{x} = Ax$.

We will not discuss nonhyperbolic cases in detail, but we make two comments. If an eigenvalue is 0, there are other equilibria besides the origin (at least a line of equilibria). If a pair of eigenvalues is pure imaginary, there are closed orbits (at least a plane of closed orbits). For example, the eigenvalues of the linear system (9.3)–(9.4) are $\pm i$. All its orbits are closed.

9.6 Linearization

Suppose $\dot{x} = f(x)$ has an equilibrium at x_0. To study solutions near x_0, we make the substitution $x = x_0 + y$. Then small y corresponds to x near x_0. We obtain $\dot{y} = f(x_0 + y)$. By Taylor's Theorem

$$\dot{y} = f(x_0) + Df(x_0)y + \cdots = Df(x_0)y + \cdots$$

because x_0 is an equilibrium. Here $Df(x_0)$ is the $n \times n$ matrix whose ij-entry is $\partial f_i / \partial x_j$ evaluated at the point x_0.

The *linearization* of the differential equation $\dot{x} = f(x)$ at the equilibrium x_0 is the linear differential equation $\dot{y} = Df(x_0)y$. We can determine the phase portrait of $\dot{y} = Df(x_0)y$ by finding eigenvalues and eigenvectors.

The equilibrium x_0 of $\dot{x} = f(x)$ is called *hyperbolic* if the linear differential equation $\dot{y} = Df(x_0)y$ is hyperbolic.

Theorem 9.3. *Linearization Theorem. If x_0 is a hyperbolic equilibrium of $\dot{x} = f(x)$, then the phase portrait of $\dot{x} = f(x)$ near x_0 looks just like the phase portrait of $\dot{y} = Df(x_0)y$ near the origin.*

The meaning of this theorem is as follows:

- If all eigenvalues of $Df(x_0)$ have negative real part, then x_0 is an asymptotically stable equilibrium of $\dot{x} = f(x)$. The equilibrium x_0 is called an *attractor*.

- If all eigenvalues of $Df(x_0)$ have positive real part, then x_0 is an asymptotically stable equilibrium of $\dot{x} = -f(x)$. In other words, for $\dot{x} = f(x)$, all solutions that start near x_0 stay near x_0 *in backward time*, and approach x_0 as $t \to -\infty$. The equilibrium x_0 is called a *repeller*.
- If $Df(x_0)$ has k eigenvalues with negative real part and $n - k$ eigenvalues with positive real part ($0 < k < n$), then there are "surfaces" W^s of dimension k and W^u of dimension $n - k$ through x_0 such that the following is true. If $x(t)$ is a solution of $\dot{x} = f(x)$ that starts near x_0, then

    — $x(t)$ stays near x_0 in forward time and approaches x_0 as $t \to \infty$ if and only if $x(0) \in W^s$;

    — $x(t)$ stays near x_0 in backward time and approaches x_0 as $t \to -\infty$ if and only if $x(0) \in W^u$.

The equilibrium x_0 is called a *saddle*. The "surfaces" W^s and W^u are called the *stable manifold* and the *unstable manifold* of x_0, respectively. (A *manifold* is a generalization of a surface. A one-dimensional manifold is a curve, and a two-dimensional manifold is a surface.) W^s and W^u are tangent at x_0 to the stable and unstable subspaces, respectively, of $\dot{y} = Df(x_0)y$, translated to x_0.

Like the existence of a first integral, this theorem is often helpful in drawing phase portraits.

Example. In the predator-prey system (9.5)–(9.6), let us add the assumption that if the prey population exceeds an environmental carrying capacity, its growth rate will become negative. A reasonable model that incorporates this assumption is

$$\dot{x} = x(a - by - \delta x), \tag{9.13}$$
$$\dot{y} = y(-c + ex), \tag{9.14}$$

with a, b, c, e, and δ positive constants.

Nullclines, equilibria, and the vector field are shown shown in Figure 9.7 (first quadrant only) under the assumption $ae - c\delta > 0$.

As in Figure 9.4, we cannot tell whether solutions that wind around the equilibrium in the interior of the first quadrant spiral in, spiral out, or close up. It is also possible that the solutions don't spiral at all; they could approach the equilibrium directly from a little north of west or a little south of east. The phase portrait near the equilibrium $(\frac{a}{\delta}, 0)$ is also unclear.

To examine the equilibria more closely, we use linearization. First, rewrite (9.5)–(9.6) a little:

$$\dot{x} = f_1(x, y) = ax - bxy - \delta x^2,$$
$$\dot{y} = f_2(x, y) = -cy + exy,$$

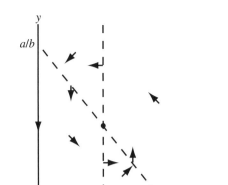

Figure 9.7. Nullclines (axes and dashed lines), equilibria (dots), and vector field of (9.13)-(9.14) in the first quadrant, assuming $ae - c\delta > 0$.

Now calculate

$$\begin{pmatrix} \dfrac{\partial f_1}{\partial x} & \dfrac{\partial f_1}{\partial y} \\ \dfrac{\partial f_2}{\partial x} & \dfrac{\partial f_2}{\partial y} \end{pmatrix} = \begin{pmatrix} a - by - 2\delta x & -bx \\ ey & -c + ex \end{pmatrix} \tag{9.15}$$

There are three equilibria, namely,

$$(0,0), \quad \left(\frac{a}{\delta}, 0\right), \quad \text{and} \quad \left(\frac{c}{e}, \frac{ae - c\delta}{be}\right).$$

At these three equilibria, the matrix (9.15) is, respectively,

$$\begin{pmatrix} a & 0 \\ 0 & -c \end{pmatrix}, \quad \begin{pmatrix} -a & -\dfrac{ab}{\delta} \\ 0 & \dfrac{ae - c\delta}{\delta} \end{pmatrix}, \quad \text{and} \quad \begin{pmatrix} -\dfrac{c\delta}{e} & -\dfrac{bc}{e} \\ \dfrac{ae - c\delta}{b} & 0 \end{pmatrix}$$

Therefore:

- At $(0,0)$ the eigenvalues are $a > 0$ and $-c < 0$. This equilibrium is a saddle.
- At $(a/\delta, 0)$ the eigenvalues are $-a < 0$ and $(ae - c\delta)/\delta > 0$. This equilibrium is also a saddle.
- At $(c/e, (ae - c\delta)/be)$ the eigenvalues turn out to be

$$-\frac{c\delta}{2e} \pm \left(\frac{c^2\delta^2}{4e^2} - \frac{c(ae - c\delta)}{e} \right)^{1/2}.$$

Both have negative real part, so this equilibrium is an attractor. For small δ both eigenvalues are complex.

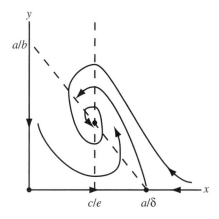

Figure 9.8. A possible phase portrait of (9.13)–(9.14), with small δ, in the first quadrant.

A phase portrait for small δ that is consistent with all our information is shown in Figure 9.8. However, more complicated phase portraits are also consistent with our information. For example, there could be a single closed orbit that surrounds the interior equilibrium. Inside the closed orbit, solutions would spiral toward the equilibrium; outside the closed orbit, solutions might spiral toward the closed orbit itself.

Chapter 10

Evolutionary dynamics

In this chapter we look at how strategies might change over time. As in our study of evolutionary stability (Sections 8.3–8.4), the context is population biology, but the ideas also apply to social dynamics.

10.1 Replicator system

As in Sections 8.3–8.4, we consider an evolutionary game based on a symmetric two-player game in normal form with finite strategy set $S = \{s_1, \ldots, s_n\}$. There is a population that uses strategy s_1 with probability p_1,...,strategy s_n with probability p_n; all $p_i \geqslant 0$ and $\sum p_i = 1$. The population state is $\sigma = \sum p_i s_i$.

When an individual of type i plays the game against a randomly chosen individual from a population with state σ, her expected payoff is $\pi_1(s_i, \sigma)$. When two randomly chosen individuals from a population with state σ play the game, the expected payoff to the first is $\pi_1(\sigma, \sigma)$.

In this chapter we explicitly regard the population state σ as changing with time. Thus we write $\sigma(t) = \sum p_i(t)s_i$.

It is reasonable to expect that if $\pi_1(s_i, \sigma) > \pi_1(\sigma, \sigma)$, then individuals using strategy i will in general have an above-average number of offspring. Thus $p_i(t)$ should increase. However, if $\pi_1(s_i, \sigma) < \pi_1(\sigma, \sigma)$, we expect $p_i(t)$ to decrease.

In fact, it is reasonable to suppose that the growth rate of p_i is proportional to $\pi_1(s_i, \sigma) - \pi_1(\sigma, \sigma)$. For simplicity we assume that the constants of proportionality are all equal to 1.

With these assumptions we obtain the *replicator system*:

$$\dot{p}_i = (\pi_1(s_i, \sigma) - \pi_1(\sigma, \sigma))p_i, \quad i = 1, \ldots, n. \tag{10.1}$$

The replicator system can also be used in social situations in which successful strategies spread, because they are seen to be successful and hence are adopted by others. In this case, it is reasonable to expect that if $\pi_1(s_i, \sigma) >$

$\pi_1(\sigma, \sigma)$, then strategy i, because it is seen to be more successful than average, will be adopted by more members of the society, so p_i will increase. If we suppose that the growth rate of p_i is proportional to $\pi_1(s_i, \sigma) - \pi_1(\sigma, \sigma)$ and that the constants of proportionality are all equal to 1, we again get the replicator system.

The replicator system is a differential equation on $\mathbb{R}^n$. The physically relevant subset of $\mathbb{R}^n$ is the *simplex*

$$\Sigma = \left\{ (p_1, \ldots, p_n) : \text{all } p_i \geqslant 0 \text{ and } \sum p_i = 1 \right\}.$$

Σ can be decomposed as follows. For each nonempty subset I of $\{1, \ldots, n\}$, let

$$\Sigma_I = \left\{ (p_1, \ldots, p_n) : p_i > 0 \text{ if } i \in I, \ p_i = 0 \text{ if } i \notin I, \text{ and } \sum p_i = 1 \right\}.$$

Then Σ is the disjoint union of the Σ_I, where I ranges over all nonempty subsets of $\{1, \ldots, n\}$. The Σ_I are called the *strata* of Σ. See Figure 10.1.

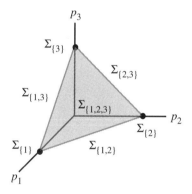

Figure 10.1. The simplex Σ with $n = 3$ and its decomposition into strata.

Let $|I|$ denote the size of the set I. The dimension of the stratum Σ_I is $|I| - 1$. For example, $\Sigma_{\{1,\ldots,n\}}$, the interior of Σ, has dimension $n - 1$, and for each $i \in \{1, \ldots, n\}$, $\Sigma_{\{i\}}$ is a point (which has dimension 0).

Theorem 10.1. *The replicator system has the following properties:*

(1) *If $p_i = 0$, then $\dot{p}_i = 0$.*
(2) *Let $S(p_1, \ldots, p_n) = \sum p_i$. If $S = 1$, then $\dot{S} = 0$.*
(3) *Each stratum Σ_I is invariant.*
(4) *Each stratum $\Sigma_{\{i\}}$ is an equilibrium.*

Proof. Property (1) follows immediately from (10.1).

To show property (2), just note that if $\sum p_i = 1$, then

$$\dot{S} = \sum \dot{p}_i = \sum (\pi_1(s_i, \sigma) - \pi_1(\sigma, \sigma)) p_i$$
$$= \sum p_i \pi_1(s_i, \sigma) - \sum p_i \pi_1(\sigma, \sigma)$$
$$= \pi_1(\sigma, \sigma) - \pi_1(\sigma, \sigma)$$
$$= 0.$$

To prove property (3), let

$$A_I = \left\{ (p_1, \ldots, p_n) : p_i = 0 \text{ if } i \notin I \text{ and } \sum p_i = 1 \right\}.$$

We have $\Sigma_I \subset A_I$, and from properties (1) and (2), A_I is invariant. Let $p(t)$ be a solution of the replicator system with $p(0) \in \Sigma_I$. Then $p(t)$ stays in A_I. Thus the only way $p(t)$ can leave Σ_I is if some $p_i(t)$, $i \in I$, becomes 0. If this happens, $p(t)$ enters A_J, where J is some proper subset of I. Since A_J is invariant, this is impossible.

To prove property (4), just note that each $\Sigma_{\{i\}}$ is a single point and is invariant, so it must be an equilibrium. $\qquad\square$

For a subset I of $\{1, \ldots, n\}$, let G_I be the reduced game derived from G by eliminating, for both players, all strategies s_i except those with $i \in I$. The *closure* of the stratum Σ_I, denoted $\mathrm{cl}(\Sigma_I)$, is just Σ_I together with all its boundary points. Equivalently, it is the union of Σ_I and all Σ_J with $J \subset I$. An important fact is that the restriction of the replicator system to $\mathrm{cl}(\Sigma_I)$ is just the replicator system for the evolutionary game based on the reduced game G_I. We will see this in examples.

Let

$$D = \left\{ (p_1, \ldots, p_{n-1}) : p_i \geqslant 0 \text{ for } i = 1, \ldots, n-1, \text{ and } \sum_{i=1}^{n-1} p_i \leqslant 1 \right\}.$$

Then

$$\Sigma = \left\{ (p_1, \ldots, p_n) : (p_1, \ldots, p_{n-1}) \in D \text{ and } p_n = 1 - \sum_{i=1}^{n-1} p_i \right\}.$$

Instead of studying the replicator system on Σ, one can instead take the space to be D and use only the first $n-1$ equations of the replicator system. In these equations, one must of course let $p_n = 1 - \sum_{i=1}^{n-1} p_i$.

For $n = 2$, the set D is simply the line segment $0 \leqslant p_1 \leqslant 1$. The endpoints $p_1 = 0$ and $p_1 = 1$ are always equilibria. Since $p_1 = 1$ is an equilibrium, $\dot{p}_1$ always has $1 - p_1$ as a factor.

For $n = 3$, the set D is the triangle $\{(p_1, p_2): p_1 \geqslant 0, \; p_2 \geqslant 0, \text{ and } p_1 + p_2 \leqslant 1\}$. The vertices $(0,0)$, $(1,0)$, and $(0,1)$ are equilibria, and the lines $p_1 = 0$, $p_2 = 0$, and $p_1 + p_2 = 1$ are invariant. Since the line $p_1 + p_2 = 1$ is invariant, we must have $\dot{p}_1 + \dot{p}_2 = 0$ whenever $p_1 + p_2 = 1$.

Notice that D is divided into strata that correspond to those of Σ.

10.2 Microsoft vs. Apple

In the early days of personal computing, people faced a dilemma. You could buy a computer running Microsoft Windows, or one running the Apple operating system. Either was reasonably satisfactory, although Apple's was better. However, neither type of computer dealt well with files produced by the other. Thus if your coworker used Windows and you used Apple, not much got accomplished.

We model this situation as a symmetric two-player game in normal form. The strategies are buy Microsoft (m) or buy Apple (a). The payoffs are given by the following matrix:

		Player 2	
		m	a
Player 1	m	$(1,1)$	$(0,0)$
	a	$(0,0)$	$(2,2)$

We are in case (3) of Theorem 8.5, a game like Stag Hunt. There are two pure-strategy strict Nash equilibria, (m, m) and (a, a). Both m and a are evolutionarily stable states. There is also a symmetric mixed strategy Nash equilibrium (σ^*, σ^*) with $\sigma^* = \frac{2}{3}m + \frac{1}{3}a$.

The Nash equilibria (m, m) and (a, a) are easy to understand intuitively. Clearly if your coworker is using Microsoft, you should use it too. Since, for each player, Microsoft is the best response to Microsoft, (m, m) is a Nash equilibrium. The same reasoning applies to (a, a).

The mixed strategy Nash equilibrium is harder to understand. One feels that it does not correspond to any behavior one would ever observe. Even if for some reason people picked computers randomly, why would they choose the worse computer with higher probability?

To resolve this mystery, we imagine a large population of people who randomly encounter each other and play this two-player game. People observe which strategy, buy Microsoft or buy Apple, is on average producing higher payoffs. They will tend to use the strategy that they observe produces the higher payoff.

Let a state of the population be $\sigma = p_1 m + p_2 a$, and let $p = (p_1, p_2)$. It is consistent with our understanding of the situation to assume that $p(t)$ evolves by the replicator system. We have

$$\pi_1(m, \sigma) = p_1, \qquad \pi_1(a, \sigma) = 2p_2,$$
$$\pi_1(\sigma, \sigma) = p_1 \pi_1(m, \sigma) + p_2 \pi_1(a, \sigma) = p_1^2 + 2p_2^2,$$

so the replicator system is

$$\dot{p}_1 = (\pi_1(m, \sigma) - \pi_1(\sigma, \sigma))p_1 = (p_1 - (p_1^2 + 2p_2^2))p_1,$$
$$\dot{p}_2 = (\pi_1(a, \sigma) - \pi_1(\sigma, \sigma))p_2 = (2p_2 - (p_1^2 + 2p_2^2))p_2.$$

Instead of drawing the phase portrait on the simplex $\Sigma = \{(p_1, p_2): p_1 \geqslant 0, \; p_2 \geqslant 0, \text{ and } p_1 + p_2 = 1\}$, we will, as explained in the Section 10.1, draw the phase portrait on $D = \{p_1 : 0 \leqslant p_1 \leqslant 1\}$. Thus we only need the first equation, in which we substitute $p_2 = 1 - p_1$:

$$\dot{p}_1 = \left(p_1 - (p_1^2 + 2(1 - p_1)^2) \right) p_1$$
$$= (1 - p_1)(p_1 - 2(1 - p_1))p_1$$
$$= (1 - p_1)(3p_1 - 2)p_1.$$

Note that we simplified the equation for $\dot{p}_1$ by factoring out $1 - p_1$ from the expression $p_1 - (p_1^2 + 2(1 - p_1)^2)$. This may seem pretty clever. However, we knew in advance, as mentioned in the previous section, that there had to be an equilibrium at $p = 1$, so $1 - p_1$ had to be a factor. To perform the factoring, we just needed to notice that $p_1 - p_1^2 = (1 - p_1)p_1$. This idea will be useful at other points in this chapter.

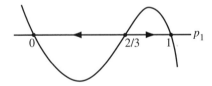

Figure 10.2. Graph of $\dot{p}_1 = (1 - p_1)(3p_1 - 2)p_1$ and phase portrait.

The phase portrait on the interval $0 \leqslant p_1 \leqslant 1$ is shown in Figure 10.2. We see that there are attracting equilibria at $p_1 = 0$ (everyone uses Apple) and $p_1 = 1$ (everyone uses Microsoft), as expected. The equilibrium at $p_1 = \frac{2}{3}$ is unstable. It separates the basin of attraction of $p_1 = 0$ from the basin of attraction of $p_1 = 1$. The location of this equilibrium now makes intuitive sense: the basin of attraction of $p_1 = 0$, in which everyone uses the better computer Apple, is larger than the basin of attraction of $p_1 = 1$, in which everyone uses the worse computer Microsoft. Nevertheless, if initially more than $\frac{2}{3}$ of the population uses Microsoft, eventually everyone uses Microsoft, even though it is worse.

10.3 Evolutionary dynamics with two pure strategies

In this Section we generalize our work on Microsoft vs. Apple. As in Section 8.4, consider an evolutionary game based on a symmetric two-player game in normal form with just two pure strategies. The payoff matrix has the form

		Player 2	
		s_1	s_2
Player 1	s_1	(a,a)	(b,c)
	s_2	(c,b)	(d,d)

A population state is $\sigma = p_1 s_1 + p_2 s_2$. We have

$$\pi_1(s_1,\sigma) = p_1 a + p_2 b, \qquad \pi_1(s_2,\sigma) = p_1 c + p_2 d,$$

$$\pi_1(\sigma,\sigma) = p_1 \pi_1(s_1,\sigma) + p_2 \pi_1(s_2,\sigma) = p_1^2 a + p_1 p_2 (b+c) + p_2^2 d.$$

Therefore the replicator system is

$$\dot{p}_1 = \big(\pi_1(s_1,\sigma) - \pi_1(\sigma,\sigma)\big)p_1$$
$$= \big(p_1 a + p_2 b - (p_1^2 a + p_1 p_2(b+c) + p_2^2 d)\big)p_1,$$
$$\dot{p}_2 = \big(\pi_1(s_2,\sigma) - \pi_1(\sigma,\sigma)\big)p_2$$
$$= \big(p_1 c + p_2 d - (p_1^2 a + p_1 p_2(b+c) + p_2^2 d)\big)p_2.$$

As explained in Section 10.1, and as with Microsoft vs. Apple, we only need the first equation, in which we substitute $p_2 = 1 - p_1$:

$$\dot{p}_1 = \big(p_1 a + (1 - p_1)b - (p_1^2 a + p_1(1 - p_1)(b+c) + (1 - p_1)^2 d)\big)p_1$$
$$= p_1(1 - p_1)(p_1 a + b - p_1(b+c) - (1 - p_1)d)$$
$$= p_1(1 - p_1)\big(b - d + ((a - c) + (d - b))p_1\big)$$
$$= p_1(1 - p_1)\big(((a - c) + (d - b))p_1 - (d - b)\big)$$

In the second line of the calculation, we moved the p_1 that was at the right to the left, and we factored out $1 - p_1$. To accomplish the factoring, we first grouped $p_1 a$ and $-p_1^2 a$. The other terms already had $1 - p_1$ as a factor. This is analogous to what we did in the Microsoft vs. Apple example (Section 10.2).

There are equilibria at $p_1 = 0$, $p_1 = 1$, and

$$p_1 = p_1^* = \frac{d - b}{(a - c) + (d - b)},$$

provided the latter is between 0 and 1. The first two factors in our final equation for $\dot{p}_1$ are positive between 0 and 1, so the sign of $\dot{p}_1$ there depends on the third factor. The phase portraits are shown in Figure 10.3. Compare Theorem 8.5. In all cases the asymptotically stable equilibria correspond to evolutionarily stable population states.

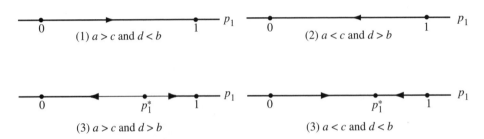

Figure 10.3. Dynamics of the replicator equation $\dot{p}_1 = p_1(1-p_1)(((a-c)+(d-b))p_1 - (d-b))$. In cases (1) and (2), there is a strictly dominant strategy; the population evolves toward everyone using it. Case (3) includes Stag Hunt and Microsoft vs. Apple. Case (4) includes Chicken.

10.4 Hawks and Doves revisited

We consider again the game of Hawks and Doves from Subsection 8.4.3. The payoff matrix is:

		Animal 2	
		h	**d**
Animal 1	**h**	$(\frac{v-w}{2}, \frac{v-w}{2})$	$(v, 0)$
	d	$(0, v)$	$(\frac{v}{2} - t, \frac{v}{2} - t)$

Recall that there are no symmetric pure-strategy Nash equilibria, and that if $p_1^* = (v + 2t)/(w + 2t)$ and $\sigma^* = p_1^* h + (1 - p_1^*)d$, then (σ^*, σ^*) is a symmetric mixed-strategy Nash equilibrium.

Now think of $\sigma = p_1 h + p_2 d$ as a population state of an evolutionary game. We recall that σ^* is an evolutionarily stable state. We have

$$\pi_1(h, \sigma) = p_1 \frac{v - w}{2} + p_2 v, \qquad \pi_1(d, \sigma) = p_2\left(\frac{v}{2} - t\right),$$

$$\pi_1(\sigma, \sigma) = p_1 \pi_1(h, \sigma) + p_2 \pi_1(d, \sigma) = p_1^2 \frac{v - w}{2} + p_1 p_2 v + p_2^2\left(\frac{v}{2} - t\right).$$

Hence the replicator system is

$$\dot{p}_1 = (\pi_1(h, \sigma) - \pi_1(\sigma, \sigma)) p_1$$
$$= \left(p_1 \frac{v - w}{2} + p_2 v - \left(p_1^2 \frac{v - w}{2} + p_1 p_2 v + p_2^2\left(\frac{v}{2} - t\right)\right)\right) p_1,$$

$$\dot{p}_2 = (\pi_1(d, \sigma) - \pi_1(\sigma, \sigma)) p_2$$
$$= \left(p_2\left(\frac{v}{2} - t\right) - \left(p_1^2 \frac{v - w}{2} + p_1 p_2 v + p_2^2\left(\frac{v}{2} - t\right)\right)\right) p_2.$$

Again we only need the first equation, in which we substitute $p_2 = 1 - p_1$:

$$\dot{p}_1 = \left(p_1 \frac{v - w}{2} + (1 - p_1)v\right.$$
$$\left. - \left(p_1^2 \frac{v - w}{2} + p_1(1 - p_1)v + (1 - p_1)^2\left(\frac{v}{2} - t\right)\right)\right) p_1$$
$$= (1 - p_1)\left(p_1 \frac{v - w}{2} + (1 - p_1)v - (1 - p_1)\left(\frac{v}{2} - t\right)\right) p_1$$
$$= (1 - p_1)\left(\frac{v}{2} + t - p_1\left(\frac{w}{2} + t\right)\right) p_1$$
$$= \tfrac{1}{2}(1 - p_1)(v + 2t - p_1(w + 2t)) p_1.$$

The phase portrait on the interval $0 \leqslant p_1 \leqslant 1$ is shown in Figure 10.4. In this case, if the population is initially anything other than $p_1 = 0$ or $p_1 = 1$, the population tends toward the state σ^*.

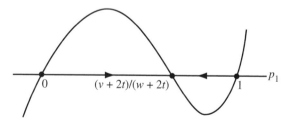

Figure 10.4. Graph of $\dot{p}_1 = \tfrac{1}{2}(1 - p_1)(v + 2t - p_1(w + 2t))p_1$ and phase portrait.

Alternatively, we could get this phase portrait by observing that we are in case (4) of Figure 10.3.

10.5 Side-blotched Lizards

The side-blotched lizard, which lives in the deserts of western North America, has three types of males:

- Orange-throats are aggressive, keep large harems of females, and defend large territories.
- Blue-throats are also aggressive, keep just one female, and defend small territories.
- Yellow-stripes are sneaky. They do not keep females or defend a territory. Instead they sneak into other males' territories and mate with their females.

Field reports indicate that populations of side-blotched lizards cycle: mostly orange-throats one generation, mostly yellow-stripes the next, mostly blue-throats the next, then back to mostly orange-throats. (For more information on side-blotched lizards, see the Wikipedia article http://en.wikipedia.org/wiki/Side-blotched_lizard.)

Let's consider a competition between two different types of male side-blotched lizards:

- Orange-throats vs. yellow-stripes. The orange-throats are unable to defend their large territories against the sneaky yellow-stripes. The yellow-stripes have the advantage.
- Yellow-stripes vs. blue-throats. The blue-throats are able to defend their small territories against the yellow-stripes. The blue-throats have the advantage.
- Blue-throats vs. orange-throats. Neither type of male bothers the other. The orange-throats, with their larger harems, produce more offspring.

This simple analysis shows why the population should cycle.

To do an analysis based on the replicator equation, consider a game in which the players are two male side-blotched lizards. Each has three possible strategies: orange-throat (O), yellow-stripe (Y), and blue-throat (B). The payoffs are 0 if both use the same strategies; otherwise, we assign a payoff of 1 or -1, respectively, to the lizard that does or does not have the advantage according to our analysis. The payoff matrix is therefore

		Lizard 2		
		O	Y	B
	O	$(0,0)$	$(-1,1)$	$(1,-1)$
Lizard 1	Y	$(1,-1)$	$(0,0)$	$(-1,1)$
	B	$(-1,1)$	$(1,-1)$	$(0,0)$

This game is symmetric. It is just Rock-Paper-Scissors (see Problem 5.12.4) in disguise. Thus it has no pure-strategy Nash equilibria and just one mixed strategy Nash equilibrium (σ^*, σ^*) with $\sigma^* = \frac{1}{3}O + \frac{1}{3}Y + \frac{1}{3}B$. However, this Nash equilibrium is not an evolutionarily stable state of the corresponding evolutionary game.

Let's calculate the replicator system of the evolutionary game. Let $\sigma = p_1 O + p_2 Y + p_3 B$. Then

$$\pi_1(O, \sigma) = -p_2 + p_3, \qquad \pi_1(Y, \sigma) = p_1 - p_3, \qquad \pi_1(B, \sigma) = -p_1 + p_2.$$

Therefore

$$
\begin{aligned}
\pi_1(\sigma, \sigma) &= p_1 \pi_1(O, \sigma) + p_2 \pi_1(Y, \sigma) + p_3 \pi_1(B, \sigma) \\
&= p_1(-p_2 + p_3) + p_2(p_1 - p_3) + p_3(-p_1 + p_2) = 0.
\end{aligned}
$$

Hence the replicator system is

$$
\begin{aligned}
\dot{p}_1 &= (\pi_1(O, \sigma) - \pi_1(\sigma, \sigma))p_1 = (-p_2 + p_3)p_1, \\
\dot{p}_2 &= (\pi_1(Y, \sigma) - \pi_1(\sigma, \sigma))p_2 = (p_1 - p_3)p_2, \\
\dot{p}_3 &= (\pi_1(B, \sigma) - \pi_1(\sigma, \sigma))p_1 = (-p_1 + p_2)p_3.
\end{aligned}
$$

We only need the first and second equations, in which we substitute $p_3 = 1 - (p_1 + p_2)$:

$$\dot{p}_1 = (1 - p_1 - 2p_2)p_1, \tag{10.2}$$

$$\dot{p}_2 = (-1 + 2p_1 + p_2)p_2. \tag{10.3}$$

The simplex Σ in $\mathbb{R}^3$ corresponds to the region

$$D = \{(p_1, p_2): p_1 \geqslant 0, \ p_2 \geqslant 0, \ p_1 + p_2 \leqslant 1\}$$

in $\mathbb{R}^2$.

Let's analyze the system (10.2)–(10.3) on D.

1. Invariance of the boundary of D. This is just a check on our work. Note that if $p_1 = 0$ then $\dot{p}_1 = 0$; if $p_2 = 0$ then $\dot{p}_2 = 0$; and if $p_1 + p_2 = 1$ then

$$
\begin{aligned}
\dot{p}_1 + \dot{p}_2 &= (1 - p_1)p_1 - 2p_2 p_1 + (-1 + p_2)p_2 + 2p_1 p_2 \\
&= (1 - p_1)p_1 + (-1 + p_2)p_2 \\
&= p_2 p_1 - p_1 p_2 \\
&= 0.
\end{aligned}
$$

2. To find all equilibria of the replicator system, we solve simultaneously the pair of equations

$$\dot{p}_1 = (1 - p_1 - 2p_2)p_1 = 0, \qquad \dot{p}_2 = (-1 + 2p_1 + p_2)p_2 = 0.$$

We find that the equilibria are $(p_1, p_2) = (0, 0)$, $(0, 1)$, $(1, 0)$, and $(\frac{1}{3}, \frac{1}{3})$.

3. Nullclines: We have $\dot p_1 = 0$ on the lines $1 - p_1 - 2p_2 = 0$ and $p_1 = 0$, and we have $\dot p_2 = 0$ on the lines $-1 + 2p_1 + p_2 = 0$ and $p_2 = 0$. See Figure 10.5.

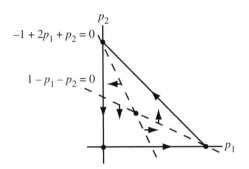

Figure 10.5. Vector field for the system (10.2)–(10.3) on D.

4. From the figure it appears that solutions circle around the equilibrium $(\frac{1}{3}, \frac{1}{3})$. We cannot, however, tell from the figure if solutions spiral toward the equilibrium, spiral away from the equilibrium, or form closed curves. It is also possible that solutions approach the equilibrium directly from a little north of west or a little south of east.

The vector field on each side of the triangle in Figure 10.5 represents the replicator equation for a reduced game in which one strategy has been eliminated. For example, on the left side of the triangle, $p_1 = 0$, so strategy 1 is missing. The payoff matrix for the reduced game is the one given earlier in this section, with the first row and first column crossed out. The remaining 2×2 matrix falls into case (2) of Figure 10.3: strategy B strictly dominates strategy Y. On the left side of the triangle, (p_1, p_2) approaches $(0, 0)$. When $p_1 = p_2 = 0$, $p_3 = 1$, that is, the population is all blue-throats.

5. We can try to get more information by linearizing the system (10.2)–(10.3) at the equilibrium $(\frac{1}{3}, \frac{1}{3})$. The linearization of (10.2)–(10.3) has the matrix

$$\begin{pmatrix} 1 - 2p_1 - 2p_2 & -2p_1 \\ 2p_2 & -1 + 2p_1 + 2p_2 \end{pmatrix}.$$

At $(p_1, p_2) = (\frac{1}{3}, \frac{1}{3})$, the matrix is

$$\begin{pmatrix} -\frac{1}{3} & -\frac{2}{3} \\ \frac{2}{3} & \frac{1}{3} \end{pmatrix}.$$

The characteristic equation is $\lambda^2 + \frac{1}{3} = 0$, so the eigenvalues are $\pm i/\sqrt{3}$. Since they are pure imaginary, the equilibrium is not hyperbolic. Thus linearization does not help. We still don't know what the solutions do.

6. Fortunately, the system (10.2)–(10.3) has the first integral $V(p_1, p_2) = \ln p_1 + \ln p_2 + \ln(1 - p_1 - p_2)$. Check:

$$
\begin{aligned}
\dot{V} &= \frac{\dot{p}_1}{p_1} + \frac{\dot{p}_2}{p_2} + \frac{-\dot{p}_1 - \dot{p}_2}{1 - p_1 - p_2} = \frac{\dot{p}_1}{p_1} + \frac{\dot{p}_2}{p_2} - \frac{\dot{p}_1 + \dot{p}_2}{1 - p_1 - p_2} \\
&= \frac{(1 - p_1 - 2p_2)p_1}{p_1} + \frac{(-1 + 2p_1 + p_2)p_2}{p_2} - \frac{(1 - p_1)p_1 + (-1 + p_2)p_2}{1 - p_1 - p_2} \\
&= (1 - p_1 - 2p_2) + (-1 + 2p_1 + p_2) - \frac{(1 - p_1 - p_2)(p_1 - p_2)}{1 - p_1 - p_2} \\
&= (1 - p_1 - 2p_2) + (-1 + 2p_1 + p_2) - (p_1 - p_2) \\
&= 0.
\end{aligned}
$$

One can check using the second derivative test that V has a local maximum at its critical point $(\frac{1}{3}, \frac{1}{3})$. Therefore level curves of $V(p_1, p_2)$ surround this point. In other words, solutions near the equilibrium $(\frac{1}{3}, \frac{1}{3})$ form closed curves around it. In fact all solutions in the interior of D form closed curves around this point.

We conclude that we expect the populations of the different types of side-blotched lizards to oscillate.

Suppose a solution of (10.2)–(10.3) returns to its initial value after time T, that is, the solution has period T. The *average* of p_i along the solution is just $\frac{1}{T} \int_0^T p_i(t)\, dt$. It turns out that on any solution, the average values of p_1 and p_2 are just their values at the interior equilibrium, namely, $\frac{1}{3}$ and $\frac{1}{3}$. We can show this by the following calculation.

Rewrite (10.2)–(10.3) as

$$
\frac{\dot{p}_1}{p_1} = 1 - p_1 - 2p_2, \tag{10.4}
$$

$$
\frac{\dot{p}_2}{p_2} = -1 + 2p_1 + p_2. \tag{10.5}
$$

Along a solution, both sides of these equations are functions of t. Integrate both sides of both equations from $t = 0$ to $t = T$:

$$
\ln p_1(T) - \ln p_1(0) = T - \int_0^T p_1(t)\, dt - 2 \int_0^T p_2(t)\, dt, \tag{10.6}
$$

$$
\ln p_2(T) - \ln p_2(0) = -T + 2 \int_0^T p_1(t)\, dt + \int_0^T p_2(t)\, dt. \tag{10.7}
$$

However, T is the period of the solution, so $\ln p_1(T) - \ln p_1(0) = 0$, and $\ln p_2(T) - \ln p_2(0) = 0$. Now it is a matter of simple algebra to show that

$$
\frac{1}{T} \int_0^T p_1(t)\, dt = \frac{1}{T} \int_0^T p_2(t)\, dt = \frac{1}{3}.
$$

10.6 Equilibria of the replicator system

In this section we derive some general facts about equilibria of the replicator system.

Theorem 10.2. *Let $p \in \Sigma$, and let σ be the corresponding population state. Then p is an equilibrium of the replicator system if and only if (σ, σ) satisfies condition (1) of the Fundamental Theorem of Nash Equilibria (Theorem 5.2).*

In other words, if $p \in \Sigma_I$ for some nonempty subset I of $\{1, \ldots, n\}$, then p is an equilibrium of the replicator system if and only if all $\pi_1(s_i, \sigma)$ with $i \in I$ are equal.

For example, if only one strategy is active at σ (i.e., one $p_i = 1$ and the others are 0), then p is automatically an equilibrium of the replicator system. Of course, we already noted this in property (4) Theorem 10.1.

Note that if all $p_i > 0$, then condition (2) of Theorem 5.2 for a Nash equilibrium is irrelevant. Hence, if all $p_i > 0$, then p is an equilibrium of the replicator system if and only if (σ, σ) is a Nash equilibrium of the game.

Proof. Let Σ_I be the stratum of Σ to which p belongs.

Suppose p is an equilibrium of the replicator system. If $i \in I$, then $p_i > 0$, so we see from (10.1) that $\pi_1(s_i, \sigma) = \pi_1(\sigma, \sigma)$. Hence all $\pi_1(s_i, \sigma)$ with $i \in I$ are equal. Thus p satisfies condition (1) of the Fundamental Theorem of Nash Equilibria.

On the other hand, suppose p satisfies condition (1) of the Fundamental Theorem of Nash Equilibria. Thus $\pi_1(s_i, \sigma) = K$ for all i such that $p_i > 0$, that is, for all $i \in I$. We have

$$\pi_1(\sigma, \sigma) = \sum_{i=1}^{n} p_i \pi_1(s_i, \sigma) = \sum_{i \in I} p_i \pi_1(s_i, \sigma) = \sum_{i \in I} p_i K = K.$$

Hence if $p_i > 0$, then $\pi_1(s_i, \sigma) = \pi_1(\sigma, \sigma)$. Now we see from (10.1) that p is an equilibrium of the replicator system. $\square$

Theorem 10.3. *Let $p^* \in \Sigma_I$ be an equilibrium of the replicator system, and let σ^* be the corresponding population state. Suppose (σ^*, σ^*) does not satisfy condition (2) of Theorem 5.2; that is, suppose there is an $i \notin I$ such that $\pi_1(s_i, \sigma^*) > \pi_1(\sigma^*, \sigma^*)$. Let J be a subset of $\{1, \ldots, n\}$ that includes both i and I. Then no solution $p(t)$ of the replicator system that lies in Σ_J approaches p^* as $t \to \infty$.*

Proof. By assumption, $\pi_1(s_i, \sigma^*) - \pi_1(\sigma^*, \sigma^*) > 0$. Hence if, σ is close to σ^*, that is, if the corresponding point p is close to the point p^*, then $\pi_1(s_i, \sigma) - \pi_1(\sigma, \sigma) > 0$. Therefore, if p is close to p^* and $p \in \Sigma_J$, so $p_i > 0$, then $\dot{p}_i = p_i(\pi_1(s_i, \sigma) - \pi_1(\sigma, \sigma))p_i$ is positive. Hence any solution $p(t)$ that lies in Σ_J and comes close to p_* has $p_i(t)$ positive and increasing when it is near p^*. Since $p_i^* = 0$, clearly $p(t)$ does not approach p^* as t increases. □

The proof of Theorem 10.3 has a nice interpretation. It says that if we introduce into the population σ^* a small number of animals using strategy i, which does better against σ^* than σ^* does against itself, then the use of strategy i in the population will increase.

Theorem 10.4. *Let $p^* \in \Sigma$. Suppose the corresponding population state σ^* is evolutionarily stable. Then p^* is an asymptotically stable equilibrium of the replicator system.*

At the end of Section 10.3, we noticed this result for most evolutionary games with two strategies.

The general proof uses the following fact, which is a consequence of the convexity of the function $\ln x$.

Lemma 10.5. *Assume*

- $x_i > 0$ *for* $i = 1, \ldots, n$.
- $p_i > 0$ *for* $i = 1, \ldots, n$.
- $\sum p_i = 1$.

Then $\ln \left(\sum p_i x_i \right) > \sum p_i \ln x_i$ *unless* $x_1 = \cdots = x_n$.

Given this fact, we shall prove Theorem 10.4 for the case in which all $p_i^* > 0$. Define a function W with domain $\Sigma_{\{1,\ldots,n\}}$ by $W(p_1, \ldots, p_n) = \sum p_i^* \ln(p_i/p_i^*)$. Then $W(p^*) = 0$. For $p \neq p^*$,

$$W(p) = \sum p_i^* \ln \frac{p_i}{p_i^*} < \ln \left(\sum p_i^* \frac{p_i}{p_i^*} \right) = \ln \left(\sum p_i \right) = \ln 1 = 0.$$

The inequality is a consequence of Lemma 10.5; since $\sum p_i = \sum p_i^* = 1$, the only way all the quotients p_i/p_i^* can be equal is if they are all 1.

Let $V = -W$. Then $V(p^*) = 0$, and, for $p \neq p^*$, $V(p) > 0$. We can write $V(p) = -\sum p_i^* (\ln p_i - \ln p_i^*)$. Then for $p \neq p^*$,

$$\dot{V} = -\sum p_i^* \frac{1}{p_i} \dot{p}_i = -\sum \frac{p_i^*}{p_i} (\pi_1(s_i, \sigma) - \pi_1(\sigma, \sigma)) p_i$$

$$= -\sum p_i^* (\pi_1(s_i, \sigma) - \pi_1(\sigma, \sigma))$$

$$= -\sum p_i^* \pi_1(s_i, \sigma) + \sum p_i^* \pi_1(\sigma, \sigma)$$

$$= -\pi_1(\sigma^*, \sigma) + \pi_1(\sigma, \sigma) < 0.$$

The last inequality follows from the assumption that σ^* is an evolutionarily stable state with all $p_i^* > 0$. In this case, for all $\sigma \neq \sigma^*$, $\pi_1(\sigma^*, \sigma) > \pi_1(\sigma, \sigma)$.

Therefore V is a strict Lyapunov function, so p^* is asymptotically stable.

10.7 Cooperators, Defectors, and Tit-for-Tatters

As far as we know, human beings have always lived in groups and cooperated in hunting and other activities. Cooperation is also observed in other species. Can evolutionary game theory help explain this behaviour?

Let's consider the symmetric version of the cooperation dilemma discussed at the end of Section 2.4. Each player can help the other, conferring a benefit $b > 0$ on the other player at a cost $a > 0$ to herself. We assume $b > a$. The players have two strategies: cooperate by helping (c), or defect by not helping (d). The payoff matrix is

		Player 2	
		c	d
Player 1	c	$(b-a, b-a)$	$(-a, b)$
	d	$(b, -a)$	$(0, 0)$

Defect is the strictly dominant strategy for both players. However, since $b > a$, the game is a prisoner's dilemma: if both cooperate, both are better off than if both defect.

If we form the replicator equation for the corresponding evolutionary game as in Section 10.3, we will see that the cooperators die out, and only defectors are left. (We are in case 2 of Figure 10.3.) Thus the benefits of cooperation alone are not sufficient to explain why it exists.

Now let us imagine that when two random players from a population meet, they play the game twice, using one of three strategies:

- c: always cooperate.
- d: always defect.
- t: tit for tat: cooperate the first time; the second time, do what the other player did the first time.

The payoff matrix for the twice-repeated game (compare Problem 3.12.6) is

		Player 2		
		c	d	t
	c	$(2b-2a, 2b-2a)$	$(-2a, 2b)$	$(2b-2a, 2b-2a)$
Player 1	d	$(2b, -2a)$	$(0,0)$	$(b, -a)$
	t	$(2b-2a, 2b-2a)$	$(-a, b)$	$(2b-2a, 2b-2a)$

We assume $b > 2a$. Then $2b - 2a > b$, so this game has, in addition to the pure-strategy Nash equilibrium (d, d), the the pure-strategy Nash equilibrium (t, t). Both are symmetric.

To simplify our study of the replicator system, we shall only consider the case $b = 3$ and $a = 1$. The payoff matrix for the twice-repeated game becomes

		Player 2		
		c	*d*	*t*
	c	$(4,4)$	$(-2,6)$	$(4,4)$
Player 1	*d*	$(6,-2)$	$(0,0)$	$(3,-1)$
	t	$(4,4)$	$(-1,3)$	$(4,4)$

This game has the symmetric Nash equilibria (d,d), $(\frac{1}{2}d + \frac{1}{2}t, \frac{1}{2}d + \frac{1}{2}t)$, and $(p_1c + p_3t, p_1c + p_3t)$ with $p_1 + p_3 = 1$ and $0 \leqslant p_1 < \frac{1}{3}$. The last family of Nash equilibria is related to the Nash equilibria found in Problem 5.12.6 with different payoffs; $p_1 = 0$ gives the Nash equilibrium (t,t) that we saw above.

The replicator system, using only p_1 and p_2, turns out to be

$$\dot{p}_1 = -2(p_1 + p_2)p_1p_2, \tag{10.8}$$

$$\dot{p}_2 = \left(-1 + 3p_1 + 3p_2 - 2p_1p_2 - 2p_2^2\right)p_2. \tag{10.9}$$

You will be asked to check this in Problem 10.13.5. We study this system on the region

$$D = \{(p_1, p_2) \colon p_1 \geqslant 0, \ p_2 \geqslant 0, \ p_1 + p_2 \leqslant 1\}.$$

1. Invariance of the boundary of D. If $p_1 = 0$ then $\dot{p}_1 = 0$, and if $p_2 = 0$ then $\dot{p}_2 = 0$. As a check on our work, we should also check that if $p_1 + p_2 = 1$, then $\dot{p}_1 + \dot{p}_2 = 0$. However, in this problem it is useful to find all numbers c such that if $p_1 + p_2 = c$ then $\dot{p}_1 + \dot{p}_2 = 0$. If $p_1 + p_2 = c$, then

$$\dot{p}_1 + \dot{p}_2 = -2(p_1 + p_2)p_1p_2 + (-1 + 3(p_1 + p_2) - 2(p_1 + p_2)p_2)p_2$$
$$= (-2cp_1 - 1 + 3c - 2cp_2)p_2$$
$$= (-2c^2 + 3c - 1)p_2$$
$$= -(2c - 1)(c - 1)p_2.$$

This expression is identically 0 if $c = \frac{1}{2}$ or $c = 1$.

2. To find the equilibria we solve simultaneously the equations $\dot{p}_1 = 0$ and $\dot{p}_2 = 0$. We find that the equililbria are $(0,1)$, $(0, \frac{1}{2})$, and the line segment consisting of points $(p_1, 0)$ with $0 \leqslant p_1 \leqslant 1$.

3. Nullclines. We have $\dot{p}_1 = 0$ on the lines $p_1 = 0$, $p_2 = 0$, and $p_1 + p_2 = 0$. We can ignore the last, since it meets D only at the origin. We have $\dot{p}_2 = 0$ on the line $p_2 = 0$, and on a curve that we will not study in detail. Note that $\dot{p}_1 < 0$ everywhere in the interior of D.

4. Linearization shows that $(0,1)$ is an attractor and $(0, \frac{1}{2})$ is a saddle. The equilibria on the line segment $(p_1, 0)$, $0 \leqslant p_1 \leqslant 1$, all have at least one eigenvalue equal to 0. The other eigenvalue is $3p_1 - 1$, so it is negative for $0 \leqslant p_1 < \frac{1}{3}$, 0 for $p_1 = \frac{1}{3}$, and positive for $\frac{1}{3} < p \leqslant 1$.

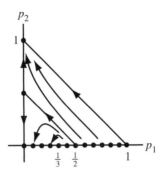

Figure 10.6. Phase portrait of the system (10.8)–(10.9).

Because of the line segment of nonhyperbolic equilibria, the phase portrait cannot be completely drawn by the methods we have learned. It is shown in Figure 10.7.

Note that the invariant line $p_1 + p_2 = \frac{1}{2}$, which we found in step 1, is the stable manifold of the saddle $(0, \frac{1}{2})$. Population states above this line are in the basin of attraction of the equilibrium $(0, 1)$, which represents a population of all defectors. These population states have $p_1 + p_2 > \frac{1}{2}$, so $p_3 = 1 - p_2 - p_2 < \frac{1}{2}$. In other words, if initially tit-for-tatters comprise less than half the population, the strategies c (cooperate) and t (tit for tat) eventually die out, and only defectors remain.

Initial conditions below the line $p_1 + p_2 = \frac{1}{2}$, which represent populations that are predominantly tit-for-tatters, lead to one of the equilibria $(p_1, 0)$ with $0 \leqslant p_1 < \frac{1}{3}$. The corresponding population states are $p_1 c + (1 - p_1)t$ with $0 \leqslant p_1 < \frac{1}{3}$. In these population states, the strategy d (defect) has died out, and only tit-for-tatters and a smaller number of cooperators remain. Apparently a large number of tit-for-tatters is required to eliminate defectors and prevent them from re-entering the population. Some members of the population can turn the other cheek should someone try to take advantage of them, but not too many! The tit-for-tatters use the strategy, do unto others as they just did unto you.

Let us compare these results to our previous theoretical work. First we discuss the reduced games that are played on the boundary of D. On the line $p_1 + p_2 = 1$, we have $p_3 = 0$, so the reduced game has no tit-for-tatters. For this reduced game, defect is the dominant strategy, so solutions tend toward all defectors. On the line $p_1 = 0$, cooperators are absent, and we have the reduced game obtained by eliminating the first row and first column of the 3×3 payoff matrix. This reduced game is of Stag Hunt type. Therefore, on the line $p_1 = 0$, a repelling equilibrium separates the basins of attraction of two attracting equilibria. Finally, on the line $p_2 = 0$, defectors are absent, and we have the reduced game obtained by eliminating the second row and second column of the 3×3 payoff matrix. In the resulting 2×2 payoff matrix, all payoffs are equal, so all strategy profiles are Nash equilibria of the reduced game. This explains the line segment of equilibria of the replicator system.

Finally, we discuss evolutionary stability. The Nash equilibria of the full game correspond to the population states d, $\frac{1}{2}d + \frac{1}{2}t$, and $p_1 c + (1 - p_1)t$, $0 \leqslant p_1 \leqslant \frac{1}{3}$. The strategy profile (d, d) is a strict Nash equilibrium, so the population state d is evolutionarily stable by Theorem 8.2, and hence the corresponding equilibrium $(0, 1)$ in D is asymptotically stable by Theorem 10.4. The population state $\frac{1}{2}d + \frac{1}{2}t$ corresponds to the equilibrium $(0, \frac{1}{2})$, which is not evolutionarily stable, because it separates equilibria of a reduced game of Stag Hunt type. None of the population states $p_1 c + (1 - p_1)t$, $0 < p_1 \leqslant \frac{1}{3}$, can be evolutionarily stable. The reason is that they all have the same active strategies; however, according to Theorem 8.4, if one were evolutionarily stable, there could be no other symmetric Nash equilibria with the same active strategies. Finally, the pure population state t is not evolutionarily stable, since it can be invaded by cooperators. Thus the only evolutionarily stable population state is d, and it corresponds to the only asymptotically stable equilibrium. However, the set of equilibria $(p_1, 0)$, $0 \leqslant p_1 < \frac{1}{3}$, enjoys a kind of asymptotic stability: for each point in the set, if an initial condition is close enough to that point, then the solution approaches the set.

10.8   Dominated strategies and the replicator system

In this section we prove two results relating iterated elimination of dominated strategies to the replicator system. Since the games we consider have two players and are symmetric, when we eliminate a strategy, we eliminate it for both players.

Theorem 10.6. *In a two-player symmetric game, suppose strategy s_i is strictly dominated by strategy s_j. Let I be a subset of $\{1, \ldots, n\}$ that contains both i and j, and let $p(t)$ be a solution of the replicator system in the stratum Σ_I of Σ. Then $p_i(t) \to 0$ as $t \to \infty$.*

Proof. Since strategy s_i is strictly dominated by strategy s_j, we have that for every pure strategy s_k, $\pi_1(s_i, s_k) < \pi_1(s_j, s_k)$. Then for any population state $\sigma = \sum p_k s_k$, we have

$$\pi_1(s_i, \sigma) = \pi_1\left(s_i, \sum p_k s_k\right) = \sum p_k \pi_1(s_i, s_k)$$
$$< \sum p_k \pi_1(s_j, s_k) = \pi_1\left(s_j, \sum p_k s_k\right) = \pi_1(s_j, \sigma).$$

Therefore, for each $p \in \Sigma$, if σ is the corresponding population state, then $\pi_1(s_i, \sigma) - \pi_1(s_j, \sigma) < 0$.

Now $\pi_1(s_i, \sigma) - \pi_1(s_j, \sigma)$ depends continuously on p, and Σ is a compact set (closed and bounded). Therefore there is a number $\epsilon > 0$ such that $\pi_1(s_i, \sigma) - \pi_1(s_j, \sigma) \leqslant -\epsilon$ for every $p \in \Sigma$.

Let $p(t)$ be a solution of the replicator system in Σ_I. Then $p_i(t) > 0$ and $p_j(t) > 0$ for all t. Therefore we can define the function $V(t) = \ln p_i(t) - \ln p_j(t)$. We have

$$
\begin{aligned}
\dot{V} &= \frac{1}{p_i}\dot{p}_i - \frac{1}{p_j}\dot{p}_j \\
&= \frac{1}{p_i}\left(\pi_1(s_i, \sigma) - \pi_1(\sigma, \sigma)\right) p_i - \frac{1}{p_j}\left(\pi_1(s_j, \sigma) - \pi_1(\sigma, \sigma)\right) p_j \\
&= \pi_1(s_i, \sigma) - \pi_1(s_j, \sigma) \leqslant -\epsilon.
\end{aligned}
$$

Then for $t > 0$,

$$
V(t) - V(0) = \int_0^t \dot{V}\, dt \leqslant \int_0^t -\epsilon\, dt = -\epsilon t.
$$

Therefore $V(t) \to -\infty$ as $t \to \infty$.

But $0 < p_j(t) < 1$, so $\ln p_j(t) < 0$, and so $-\ln p_j(t) > 0$. Since $V(t) = \ln p_i(t) - \ln p_j(t)$ approaches $-\infty$ and $-\ln p_j(t)$ is positive, it must be that $\ln p_i(t)$ approaches $-\infty$. But then $p_i(t)$ approaches 0. $\qquad\square$

We note that Theorem 10.6 does not hold when one strategy weakly dominates another. For example, in the game of Cooperators, Defectors, and Tit-for-Tatters in the Section 10.7, tit for tat weakly dominates cooperate, but solutions in the interior of the simplex do not necessarily lead to cooperate dying out.

Theorem 10.7. *In a two-player symmetric game, suppose that when we do iterated elimination of strictly dominated strategies, the strategy s_k is eliminated at some point. Let $p(t)$ be a solution of the replicator system in $\Sigma_{\{1,\ldots,n\}}$. Then $p_k(t) \to 0$ as $t \to \infty$.*

Proof. We just give the proof for the case in which only one strategy is eliminated before s_k. Let that strategy be s_i, eliminated because it is strictly dominated by a strategy s_j. Then s_k is strictly dominated by some strategy s_l once s_i *is eliminated*. This means that $\pi_1(s_k, s_m) < \pi_1(s_l, s_m)$ for every m other than i.

Let $\tilde{\Sigma}$ denote the subset of Σ on which $p_i = 0$. Then for every $p \in \tilde{\Sigma}$, $\pi_1(s_k, \sigma) < \pi_1(s_l, \sigma)$. Since $\tilde{\Sigma}$ is compact, there is a number $\epsilon > 0$ such that $\pi_1(s_k, \sigma) - \pi_1(s_l, \sigma) \leqslant -\epsilon$ for all $p \in \tilde{\Sigma}$.

By continuity, there is a number $\delta > 0$ such that if $p \in \Sigma$ and $0 \leqslant p_i < \delta$, then $\pi_1(s_k, \sigma) - \pi_1(s_l, \sigma) \leqslant -\frac{\epsilon}{2}$.

Let $p(t)$ be a solution of the replicator system in $\Sigma_{\{1,...,n\}}$. By the Theorem 10.6, $p_i(t) \to 0$ as $t \to \infty$. Therefore, for t greater than or equal to some t_0, $0 < p_i(t) < \delta$. Let $V(t) = \ln p_k(t) - \ln p_l(t)$. Then for $t \geqslant t_0$,

$$\dot{V} = \frac{1}{p_k}\dot{p}_k - \frac{1}{p_l}\dot{p}_l = \frac{1}{p_k}(\pi_1(s_k, \sigma) - \pi_1(\sigma, \sigma))p_k$$

$$- \frac{1}{p_l}(\pi_1(s_l, \sigma) - \pi_1(\sigma, \sigma))p_l$$

$$= \pi_1(s_k, \sigma) - \pi_1(s_l, \sigma) \leqslant -\frac{\epsilon}{2}.$$

Therefore, for $t > t_0$,

$$V(t) - V(t_0) = \int_{t_0}^{t} \dot{V}\, dt \leqslant \int_{t_0}^{t} -\frac{\epsilon}{2}\, dt = -\frac{\epsilon}{2}t.$$

Therefore $V(t) \to -\infty$ as $t \to \infty$.

As in the proof of the previous theorem, we can conclude that $p_k(t) \to 0$ as $t \to \infty$. $\qquad\square$

10.9 Asymmetric evolutionary games

Consider an asymmetric two-player game G in normal form. Player 1 has the finite strategy set $S = \{s_1, \ldots, s_n\}$. Player 2 has the finite strategy set $T = \{t_1, \ldots, t_m\}$. If Player 1 uses the pure strategy s_i and Player 2 uses the pure strategy t_j, the payoff to Player 1 is $\pi_1(s_i, t_j)$, and the payoff to Player 2 is $\pi_2(s_i, t_j)$.

Suppose there are two populations, one consisting of individuals like Player 1, the other consisting of individuals like Player 2. When an individual from the first population encounters an individual from the second population, they play the game G.

Taken as a whole, the first population uses strategy s_1 with probability $p_1, \ldots$, strategy s_n with probability p_n; all $p_i \geqslant 0$ and $\sum p_i = 1$. Let $\sigma = \sum p_i s_i$ be the state of the first population. Similarly, taken as a whole, the second population uses strategy t_1 with probability $q_1, \ldots$, strategy t_m with probability q_m; all $q_j \geqslant 0$ and $\sum q_j = 1$. Let $\tau = \sum q_j t_j$ be the state of the second population.

When an individual of type i from the first population plays the game against a randomly chosen individual from the second population, whose state is τ, her expected payoff is that of an individual using strategy s_i against one using the mixed strategy τ, namely,

$$\pi_1(s_i, \tau) = \sum_{j=1}^{m} q_j \pi_1(s_i, t_j).$$

Similarly, when an individual of type j from the second population plays the game against a randomly chosen individual from the first population, whose state is σ, her expected payoff is

$$\pi_2(\sigma, t_j) = \sum_{i=1}^{n} p_i \pi_2(s_i, t_j).$$

When two randomly chosen individuals from the two populations play the game, the expected payoff to the first is

$$\pi_1(\sigma, \tau) = \sum_{i=1}^{n} p_i \pi_1(s_i, \tau) = \sum_{i=1}^{n} \sum_{j=1}^{m} p_i q_j \pi_1(s_i, t_j).$$

Similarly, the expected payoff to the second is

$$\pi_2(\sigma, \tau) = \sum_{j=1}^{m} q_j \pi_2(\sigma, t_j) = \sum_{j=1}^{m} \sum_{i=1}^{n} q_j p_i \pi_2(s_i, t_j) = \sum_{i=1}^{n} \sum_{j=1}^{m} p_i q_j \pi_2(s_i, t_j).$$

We combine the two population states σ and τ into a total population state (σ, τ), and regard (σ, τ) as changing with time. Reasoning as in Section 10.1, we obtain the replicator system:

$$\dot{p}_i = (\pi_1(s_i, \tau) - \pi_1(\sigma, \tau))p_i, \quad i = 1, \dots, n; \tag{10.10}$$
$$\dot{q}_j = (\pi_2(\sigma, t_j) - \pi_2(\sigma, \tau))q_j, \quad j = 1, \dots, m. \tag{10.11}$$

Let

$$\Sigma_n = \left\{ (p_1, \dots, p_n) : \text{all } p_i \geq 0 \text{ and } \sum p_i = 1 \right\},$$
$$\Sigma_m = \left\{ (q_1, \dots, q_m) : \text{all } q_j \geq 0 \text{ and } \sum q_j = 1 \right\}.$$

The system (10.10)–(10.11) should be considered on $\Sigma_n \times \Sigma_m$.
Let

$$D_{n-1} = \left\{ (p_1, \dots, p_{n-1}) : p_i \geq 0 \text{ for } i = 1, \dots, n-1, \text{ and } \sum_{i=1}^{n-1} p_i \leq 1 \right\},$$
$$D_{m-1} = \left\{ (q_1, \dots, q_{m-1}) : q_j \geq 0 \text{ for } j = 1, \dots, m-1, \text{ and } \sum_{j=1}^{m-1} q_j \leq 1 \right\}.$$

Instead of studying an asymmetric replicator system on $\Sigma_n \times \Sigma_m$, one can instead take the space to be $D_{n-1} \times D_{m-1}$, and use only the differential equations for $\dot{p}_1, \dots, \dot{p}_{n-1}$ and $\dot{q}_1, \dots, \dot{q}_{m-1}$. In these equations, one must of course let $p_n = 1 - \sum_{i=1}^{n-1} p_i$ and $q_m = 1 - \sum_{j=1}^{m-1} q_j$.

A total population state (σ^*, τ^*) is a Nash equilibrium provided

$$\pi_1(\sigma^*, \tau^*) \geqslant \pi_1(\sigma, \tau^*) \quad \text{for all } \sigma,$$
$$\pi_2(\sigma^*, \tau^*) \geqslant \pi_2(\sigma^*, \tau) \quad \text{for all } \tau.$$

However, the notion of evolutionarily stable state for symmetric games does not have an analog for asymmetric games, since individuals from the same population cannot play the game against each other.

Many results about the replicator system for symmetric games also hold for the replicator system for asymmetric games:

(1) A population state (σ, τ) is an equilibrium of the replicator system if and only if it satisfies the equality conditions for a Nash equilibrium.
(2) A point on the boundary of $\Sigma_n \times \Sigma_m$ that satisfies the equality conditions for a Nash equilibrium, but does not satisfy one of the inequality conditions, attracts no solution in which the strategy corresponding to the unsatisfied inequality condition is present.
(3) If a strategy is eliminated in the course of iterated elimination of strictly dominated strategies, then for any solution in the interior of $\Sigma_n \times \Sigma_m$, that strategy dies out.

An important difference, however, is that for asymmetric replicator systems, it is known that equilibria in the interior of $\Sigma_n \times \Sigma_m$ are never asymptotically stable [8].

10.10   Big Monkey and Little Monkey 7

As in Section 3.1, suppose Big Monkey and Little Monkey decide simultaneously whether to wait or climb. We have a game in normal form with the following payoff matrix, repeated from Section 3.1, except that we have changed the order of climb and wait:

			Little Monkey q_1 climb	q_2 wait
Big Monkey	p_1	climb	$(5, 3)$	$(4, 4)$
	p_2	wait	$(9, 1)$	$(0, 0)$

We now imagine a population of Big Monkeys and a population of Little Monkeys. Let $s_1 = t_1 = $ climb, $s_2 = t_2 = $ wait, $\sigma = p_1 s_1 + p_2 s_2$, $\tau = q_1 t_1 + q_2 t_2$, so $p_2 = 1 - p_1$ and $q_2 = 1 - q_1$. The monkeys randomly encounter a monkey of the other type and play the game.

We could write differential equations for $\dot{p}_1$, $\dot{p}_2$, $\dot{q}_1$, and $\dot{q}_2$, but we only need those for $\dot{p}_1$ and $\dot{q}_1$, so we will omit the other two. Using $p_2 = 1 - p_1$ and $q_2 = 1 - q_1$, we obtain

$$\begin{aligned}
\dot{p}_1 &= (\pi_1(s_1,\tau) - \pi_1(\sigma,\tau))p_1 \\
&= \big((5q_1 + 4(1-q_1)) - (5p_1q_1 + 4p_1(1-q_1) + 9(1-p_1)q_1)\big)p_1 \\
&= p_1(1-p_1)(5q_1 + 4(1-q_1) - 9q_1) = p_1(1-p_1)(4 - 8q_1), \\
\dot{q}_1 &= (\pi_2(\sigma,t_1) - \pi_2(\sigma,\tau))q_1 \\
&= \big((3p_1 + 1(1-p_1)) - (3p_1q_1 + 4p_1(1-q_1) + 1(1-p_1)q_1)\big)q_1 \\
&= q_1(1-q_1)(3p_1 + (1-p_1) - 4p_1) = q_1(1-q_1)(1 - 2p_1).
\end{aligned}$$

We consider this system on

$$D_1 \times D_1 = \{(p_1,q_1)\colon 0 \leqslant p_1 \leqslant 1 \text{ and } 0 \leqslant q_1 \leqslant 1\}.$$

See Figure 10.7

1. Invariance of the boundary of $D_1 \times D_1$. This is just a check on our work. Note that if $p_1 = 0$ or $p_1 = 1$, then $\dot{p}_1 = 0$; and if $q_1 = 0$ or $q_1 = 1$, then $\dot{q}_1 = 0$.

2. To find all equilibria of the replicator system, we solve simultaneously the pair of equations $\dot{p}_1 = 0$ and $\dot{q}_1 = 0$. We find that the equilibria are $(p_1,q_1) = (0,0)$, $(0,1)$, $(1,0)$, $(1,1)$, and $(\frac{1}{2},\frac{1}{2})$.

3. Nullclines: We have $\dot{p}_1 = 0$ on the lines $p_1 = 0$, $p_1 = 1$, and $q_1 = \frac{1}{2}$; and we have $\dot{q}_1 = 0$ on the lines $q_1 = 0$, $q_1 = 1$, and $p_1 = \frac{1}{2}$. See Figure 10.7.

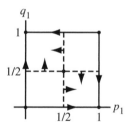

Figure 10.7. Vector field for the evolving monkeys.

4. It appears that the corner equilibria are attractors or repellers and the interior equilibrium is a saddle. This is correct and can be checked by linearization. The phase portrait is given in Figure 10.8.

The stable manifold of the saddle separates the basin of attraction of the point $(0,1)$, where Big Monkey waits and Little Monkey climbs, from the basin of attraction of the point $(1,0)$, where Big Monkey climbs and Little Monkey

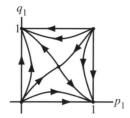

Figure 10.8. Phase portrait for the evolving monkeys.

waits. Thus we expect to observe the population in one or the other of these pure states, but we can't guess which without knowing where the population started.

Each side of the square represents a reduced game in which one of the monkeys has chosen a definite strategy. The other monkey's strategy evolves toward its best response.

10.11   Hawks and Doves with Unequal Value

Two animal species dispute a food source, but the source is more valuable to Species 2 than to Species 1. When an animal from Species 1 meets an animal from Species 2 near the food source, each has the two possible strategies Hawk and Dove from the game of Hawks and Doves in Subsection 8.4.3. The payoffs are given by the following matrix:

		Species 2	
		h	d
Species 1	h	$(\frac{v_1-w}{2}, \frac{v_2-w}{2})$	$(v_1, 0)$
	d	$(0, v_2)$	$(\frac{v_1}{2}, \frac{v_2}{2})$

In this matrix , v_i is the value of the food source to an animal of Species i. The cost of injury w is assumed to be the same for both species. We assume $v_1 < v_2 < w$. For simplicity we have taken the cost of protracted display t to be 0.

There are two pure strategy Nash equilibria, (h, d) and (d, h), and a mixed strategy Nash equilibrium

$$\left(\frac{v_2}{w}h + \left(1 - \frac{v_2}{w}\right)d, \frac{v_1}{w}h + \left(1 - \frac{v_1}{w}\right)d\right).$$

We now imagine populations of Species 1 and Species 2. Let $\sigma = p_1h_1 + p_2d$, $\tau = q_1h + q_2d$, so $p_2 = 1 - p_1$ and $q_2 = 1 - q_1$. Animals of Species 1 randomly encounter animals of Species 2 near the food source and play the game.

As in Section 10.10, we set $p_2 = 1 - p_1$ and $q_2 = 1 - q_1$, and only write the differential equations for p_1 and p_2:

$$\dot{p}_1 = (\pi_1(h, \tau) - \pi_1(\sigma, \tau))p_1$$

$$= \left(\frac{v_1 - w}{2}q_1 + v_1(1 - q_1)\right.$$

$$\left. - \left(\frac{v_1 - w}{2}p_1 q_1 + v_1 p_1(1 - q_1) + \frac{v_1}{2}(1 - p_1)(1 - q_1)\right)\right) p_1$$

$$= p_1(1 - p_1)\left(\frac{v_1}{2} - \frac{w}{2}q_1\right),$$

$$\dot{q}_1 = (\pi_2(\sigma, h) - \pi_2(\sigma, \tau))q_1$$

$$= \left(\frac{v_2 - w}{2}p_1 + v_2(1 - p_1)\right.$$

$$\left. - \left(\frac{v_2 - w}{2}p_1 q_1 + v_2(1 - p_1)q_1 + \frac{v_2}{2}(1 - p_1)(1 - q_1)\right)\right) q_1$$

$$= q_1(1 - q_1)\left(\frac{v_2}{2} - \frac{w}{2}p_1\right).$$

We consider this system on

$$D_1 \times D_1 = \{(p_1, q_1): 0 \leqslant p_1 \leqslant 1 \text{ and } 0 \leqslant q_1 \leqslant 1\}.$$

The nullclines and the vector field on the nullclines are as in Figure 10.7, except that two nullclines are $p_1 = \frac{v_2}{w}$ and $q_1 = \frac{v_1}{w}$ instead of $p_1 = \frac{1}{2}$ and $q_1 = \frac{1}{2}$. The phase portrait is shown in Figure 10.9 under the assumptions that $\frac{v_1}{w}$ is close to 0 and $\frac{v_2}{w}$ is close to 1. We see that the basin of attraction of the equilibrium $(0, 1)$ is much larger than the basin of attraction of the equilibrium $(1, 0)$. At the first equilibrium, Species 2, which places more value on the resource than does Species 1, is willing to fight for it, but Species 1 is not. The second equilibrium is the reverse.

10.12 The Ultimatum Minigame revisited

In the Ultimatum Minigame of Section 5.6, we now imagine a population of Alices (the offerers) and a population of Bobs (the responders). Let $\sigma = p_1 f + p_2 u$, $\tau = q_1 a + q_2 r$, so $p_2 = 1 - p_1$ and $q_2 = 1 - q_1$. The Bobs and Alices randomly encounter each other and play the Ultimatum Minigame. The differential equations for $\dot{p}_1$ and $\dot{q}_1$, after simplification and factoring, are

$$\dot{p}_1 = p_1(1 - p_1)(2 - 3q_1),$$
$$\dot{q}_1 = q_1(1 - q_1)(1 - p_1).$$

You should check this result.

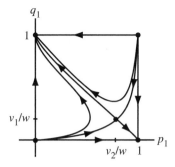

Figure 10.9. Phase portrait for Hawks and Doves with Unequal Value.

The equilibria in $D_1 \times D_1 = \{(p_1, q_1): 0 \leqslant p_1 \leqslant 1 \text{ and } 0 \leqslant q_1 \leqslant 1\}$ are $(0,0)$, $(0,1)$, and the line segment $p_1 = 1$, $0 \leqslant q_1 \leqslant 1$. Those that correspond to Nash equilibria of the game are $(0,1)$ and the smaller line segment $p_1 = 1$, $0 \leqslant q_1 \leqslant \frac{2}{3}$. Linearization shows that $(0,0)$ is a repeller; $(0,1)$ is an attractor; and points on the line segment $p_1 = 1$, $0 \leqslant q_1 \leqslant 1$ have one zero eigenvalue and one negative eigenvalue if $0 \leqslant q_1 < \frac{2}{3}$, two zero eigenvalues if $q_1 = \frac{2}{3}$, and one zero eigenvalue and one positive eigenvalue if $\frac{2}{3} < q_1 \leqslant 1$.

As in Section 10.7, because of the nonhyperbolic equilibria, the phase portrait cannot be completely drawn by the methods we have learned. It is given in Figure 10.10.

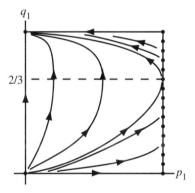

Figure 10.10. Phase portrait for the Ultimatum Minigame. The line $q_1 = \frac{2}{3}$ is a nullcline on which $\dot{p}_1 = 0$; $\dot{q}_1$ is positive everywhere in the interior of the square $D_1 \times D_1$.

From the phase portrait we see that most initial population states are in the basin of attraction of the equilibrium $(0,1)$, which corresponds to the Nash equilibrium of the game in which Alice makes an unfair offer and Bob accepts it. Recall that this is the equilibrium we expect from backward induction. Nevertheless, a sizeable set of initial conditions lead to an equilibrium $(1, q_1)$

with $0 \leqslant q_1 \leqslant \frac{2}{3}$. These are Nash equilibria in which Bob threatens to reject unfair offers with probability $1 - q_1 \geqslant \frac{1}{3}$. Alice, believing the threat, makes a fair offer. For the threat to be believable, it must be that unfair offers really do get rejected much of the time.

We conclude that it is possible for the dynamic of evolution to lead to a situation in which unfair offers are rejected with high probability and hence are not made. As mentioned in Section 5.6, experiments with the Ultimatum Game indicate that this is in fact the case.

Notice, however, that the initial conditions that lead to an equilibrium in which unfair offers are not made all have $p_1 > q_1$; that is, from the start, the fraction of Alices making fair offers is greater than the fraction of Bobs willing to accept unfair offers. You should also notice that nothing holds the population at a specific equilibrium in the set of equilibria $(1, q_1)$ with $0 \leqslant q_1 \leqslant \frac{2}{3}$. If some random fluctuation takes the population state off this set (i.e., if some Alices try the strategy u), the population state can return to a different equilibrium on the set. Thus the population state can drift along the line of equilibria. Should q_1 drift above $\frac{2}{3}$, and some Alices start to use the strategy u, the solution will go toward the stable equilibrium $(0, 1)$. For this reason, this model is generally not considered to fully explain how fairness establishes itself in populations.

Somewhat similar comments apply to the model of evolution of cooperation in Section 10.7.

10.13 Problems

10.13.1 Lions and Impalas 2.
This problem is a continuation of Problem 8.6.2.

(1) Denote a population state by $\sigma = p_1 b + p_2 l$. Find the replicator system for this game. Answer:

$$\dot{p}_1 = \Big(2p_1 + 4p_2 - (p_1(2p_1 + 4p_2) + p_2(2p_1 + p_2))\Big)p_1,$$
$$\dot{p}_2 = \Big(2p_1 + p_2 - (p_1(2p_1 + 4p_2) + p_2(2p_1 + p_2))\Big)p_2.$$

(2) Use $p_2 = 1 - p_1$ to reduce this system of two differential equations to one differential equation in the variable p_1 only. Answer:

$$\dot{p}_1 = 3p_1(1 - p_1)^2.$$

(3) Sketch the phase portrait on the interval $0 \leqslant p_1 \leqslant 1$, and describe in words what happens. (Figure 10.3 does not help, because $a = c$.)

10.13.2 Stag Hunt with Easily Spooked Hares 2. This problem is a continuation of Problem 8.6.3.

(1) Derive the replicator system and reduce it to a single differential equation.
(2) Find all equilibria of your differential equation.
(3) Draw the phase portrait of your differential equation. (Figure 10.3 does not help because $a = c$.) What does the phase portrait tell you about stag hunting and hare hunting in this problem?

10.13.3 Pure coordination games. Microsoft vs. Apple (Section 10.2) is a pure coordination game. We saw that the basin of attraction of the better attracting equilibrium is larger than the basin of attraction of the worse attracting equilibrium. Is this always true?

To explore this question, in Section 10.3 assume $a > c$ and $d > b$, so that we have a pure coordination game. The phase portrait is given by part (3) of Figure 10.3. The equilibrium $p_1 = 0$ corresponds to the pure population state s_2, and the equilibrium $p_1 = 1$ corresponds to the pure population state s_1. Assume $a > d$, so that the equilibrium $p_1 = 1$ gives the players a higher payoff than the equilibrium $p_1 = 0$.

(1) Explain the following: $p_1^* < \frac{1}{2}$ if and only if $a - c > d - b$.
(2) Explain in words under what circumstances the equilibrium $p_1 = 1$ has a larger basin of attraction than the equilibrium $p_1 = 0$.

10.13.4 Generalized Rock-Paper-Scissors. Consider a generalization of Rock-Paper-Scissors in which two players who use the same strategy receive a payoff α rather than 0. The payoff matrix is then

		Player 2		
		R	P	S
	R	(α, α)	$(-1, 1)$	$(1, -1)$
Player 1	P	$(1, -1)$	(α, α)	$(-1, 1)$
	S	$(-1, 1)$	$(1, -1)$	(α, α)

We assume that $-1 < \alpha < 1$ and $\alpha \neq 0$.

(1) Show that there are no pure strategy Nash equilibria.
(2) There is one mixed strategy Nash equilibrium, $p_1 = p_2 = p_3 = \frac{1}{3}$. (You don't have to check this.)

(3) Find the replicator system for this game. Answer:

$$\dot{p}_1 = (\alpha p_1 - p_2 + p_3 - \alpha(p_1^2 + p_2^2 + p_3^2))p_1,$$
$$\dot{p}_2 = (p_1 + \alpha p_2 - p_3 - \alpha(p_1^2 + p_2^2 + p_3^2))p_2,$$
$$\dot{p}_3 = (-p_1 + p_2 + \alpha p_3 - \alpha(p_1^2 + p_2^2 + p_3^2))p_3.$$

(4) Use $p_3 = 1 - p_1 - p_2$ to reduce this system of three differential equations to two differential equation in the variables p_1 and p_2 only. Answer:

$$\dot{p}_1 = \Big(1 + (\alpha - 1)p_1 - 2p_2 - \alpha(p_1^2 + p_2^2 + (1 - p_1 - p_2)^2)\Big)p_1,$$
$$\dot{p}_2 = \Big(-1 + 2p_1 + (\alpha + 1)p_2 - \alpha(p_1^2 + p_2^2 + (1 - p_1 - p_2)^2)\Big)p_2.$$

(5) In the region $p_1 \geq 0$, $p_2 \geq 0$, $p_1 + p_2 \leq 1$, the only equilibria are the corners and $(\frac{1}{3}, \frac{1}{3})$. The phase portrait on the boundary of the triangle is the same as in Figure 10.5. To get some idea of the phase portrait in the interior of the triangle, calculate the eigenvalues of the linearization at $(\frac{1}{3}, \frac{1}{3})$. Answer: $(\alpha \pm i\sqrt{3})/3$. What does this tell you?

10.13.5 Cooperators, Defectors, and Tit-for-Tatters. This problem is about Section 10.7.

(1) Derive the replicator system (10.8)–(10.9).
(2) Use linearization to find the eigenvalues of the equilibria on the line segment $(p_1, 0)$, $0 \leq p_1 \leq 1$.

10.13.6 Honesty and Trust Come and Go. Consider an asymmetric evolutionary game with populations of sellers and buyers. Sellers can be honest (H) or dishonest (D). Buyers can carefully inspect the merchandise (I) or can trust that it is as represented (T). The payoffs are given by the following matrix:

		Buyer	
		I	*T*
Seller	*H*	(2,3)	(3,4)
	D	(1,2)	(4,1)

Notice that the best response to an honest seller is to trust her (avoiding the time and trouble of inspecting), while the best response to dishonest seller is to inspect; and the best response to an inspecting buyer is be honest, while the best response to a trusting buyer is to be dishonest. Thus there are no pure-strategy Nash equilibria. There is a mixed-strategy Nash equilibrium in which each player uses each strategy half the time.

(1) Find the replicator system using only the variables p_1 and q_1. Answer:
$\dot{p}_1 = p_1(1 - p_1)(2q_1 - 1)$, $\dot{q}_1 = q_1(1 - q_1)(1 - 2p_1)$.

(2) In the square $\{(p_1, q_1): 0 \leqslant p_1 \leqslant 1, 0 \leqslant q_1 \leqslant 1\}$, draw the nullclines, the equilibria, and the vector field on the nullclines.

(3) From you picture, it should appear that the corner equilibria are all saddles; you don't need to check this. It should also appear that solutions spiral around the equilibrium $(\frac{1}{2}, \frac{1}{2})$ in the clockwise direction. Calculate the eigenvalues of the linearization at the equilibrium $(\frac{1}{2}, \frac{1}{2})$.

(4) You should find that the eigenvalues at the equilibrium $(\frac{1}{2}, \frac{1}{2})$ are pure imaginary. Thus Theorem 9.3 does not help us. Use the method of Subsection 9.4.2 to find a first integral. One answer: $V(p_1, q_1) = (p_1 - p_1^2)(q_1 - q_1^2)$.

(5) Show that V has a local maximum at $(\frac{1}{2}, \frac{1}{2})$. Therefore, as in Subsection 9.4.2, near the equilibrium, solutions go around the equilibrium and rejoin themselves: they are time periodic. (Actually, all solutions in the interior of the square do this.)

According to this problem, trust and honesty cycle: if sellers are generally honest, buyers start to trust, so sellers become dishonest, so buyers start to inspect, so sellers again become honest, and the cycle repeats. Similar cycles occur in the relationship between government regulators (which are like the buyers) and businesses (which are like the sellers).

10.13.7 Cooperators, Defectors, and Punishers. Does punishing defectors encourage cooperation? Sections 10.7 and 10.12 both address this question. Here is another simple model that uses an asymmetric game.

Alice can help Bob, conferring a benefit b on Bob at a cost a to herself. If Alice does not help, Bob can punish her, causing a loss l to Alice at a cost e to himself. Alice has two strategies: cooperate by helping (c) or defect by not helping (d). Bob also has two strategies: punish Alice if she defects (p), or don't bother punishing her if she defects (n). The payoff matrix is

		Bob	
		p	n
Alice	c	$(-a, b)$	$(-a, b)$
	d	$(-l, -e)$	$(0, 0)$

We shall assume $0 < a < l$ and $0 < e < b$.

(1) Use iterative elimination of weakly dominated strategies to find a dominated strategy equilibrium.

(2) Find a second pure strategy Nash equilibrium.

(3) Find the replicator system using only the variables p_1 and q_1. Answer: $\dot{p}_1 = p_1(1 - p_1)(lq_1 - a)$, $\dot{q}_1 = -e(1 - p)q(1 - q)$.

(4) Use the nullclines and linearization to draw the phase portrait. Answer: like Figure 10.10 turned upside down, with the horizontal nullcline at $q_1 = \frac{a}{l}$. Note that one of our assumptions implies that $0 < \frac{a}{l} < 1$.

10.13.8 Reputation. In this problem we explore how the previous model of Cooperators, Defectors, and Punishers changes if Bob's reputation influences Alice.

First suppose Alice is a cooperator. If she knows that Bob is a punisher, her action is not affected: she still prefers to cooperate. However, if she knows that Bob is a nonpunisher, we assume that she will prefer to defect. We suppose that when Bob is a nonpunisher, Alice realizes it with probability μ (Bob's reputation precedes him) and defects.

Next suppose Alice is a defector. If she knows that Bob is a nonpunisher, her action is not affected: she still prefers to defect. However, if she knows that Bob is a punisher, we assume that she will prefer to cooperate. We suppose that when Bob is a punisher, Alice realizes it with probability ν and cooperates.

The payoff matrix is now

		Bob p	n
Alice	c	$(-a, b)$	$(-a(1-\mu), b(1-\mu))$
	d	$(-l(1-\nu) - a\nu, -e(1-\nu) + b\nu)$	$(0, 0)$

As in the previous problem, we assume $0 < a < l$ and $0 < e < b$.

(1) Find the replicator system using only the variables p_1 and q_1. Answer:
$$\dot{p}_1 = p_1(1 - p_1)\big((l - a\mu + (a - l)\nu)q_1 - a(1 - \mu)\big),$$
$$\dot{q}_1 = \big(-e + (b + e)\nu + (e + b\mu + (b - e)\nu)p_1\big)q_1(1 - q_1).$$

(2) For $\nu = 0$ (punishers have no reputation), the replicator system becomes
$$\dot{p}_1 = p_1(1 - p_1)((l - a\mu)q_1 - a(1 - \mu)),$$
$$\dot{q}_1 = (-e + (e + b\mu)p_1)q_1(1 - q_1).$$

Use the nullclines and linearization to draw the phase portrait. Answer: like Figure 10.8 flipped across a vertical line, with the horizontal nullcline at $q_1 = (a - a\mu)/(l - a\mu)$ and the vertical nullcline at $p_1 = e/(e + b\mu)$.

(3) In the phase portrait you just drew, what are the stable equilibria? As μ increases from 0 to 1, which basin of attraction grows and which shrinks?

10.13.9 Asymmetric evolutionary games with two strategies. In this problem we look at the general asymmetric evolutionary game with two strategies. The payoff matrix is

		Player 2	
		t_1	t_2
Player 1	s_1	$(\alpha_{11}, \beta_{11})$	$(\alpha_{12}, \beta_{12})$
	s_2	$(\alpha_{21}, \beta_{21})$	$(\alpha_{22}, \beta_{22})$

(1) Show that if we replace α_{11} and α_{21} by $\alpha_{11} + k$ and $\alpha_{21} + k$, the replicator system is unchanged. Why does this make intuitive sense? The same is true if we add a constant to α_{12} and α_{22}, or to β_{11} and β_{12}, or to β_{21} and β_{22}.

(2) Because of the previous problem, we can assume that $\alpha_{11} = \beta_{11} = \alpha_{22} = \beta_{22} = 0$. We write the new payoff matrix as

		Player 2	
		t_1	t_2
Player 1	s_1	$(0,0)$	(a,b)
	s_2	(c,d)	$(0,0)$

Show that the replicator system, using only the variables p_1 and q_1, is

$$\dot{p}_1 = p_1(1 - p_1)(a - (a + c)q_1),$$
$$\dot{q}_1 = (d - (b + d)p_1)q_1(1 - q_1).$$

(3) Assume that $0 < a/(a + c) < 1$ and $0 < d/(b + d) < 1$. Then there is an interior equilibrium at $(p_1^*, q_1^*) = (d/(b + d), a/(a + c))$. Show that the linearized replicator system at (p_1^*, q_1^*) has the matrix

$$\begin{pmatrix} 0 & -(a + c)p_1^*(1 - p_1^*) \\ -(b + d)q_1^*(1 - q_1^*) & 0 \end{pmatrix}.$$

Thus the characteristic equation is

$$\lambda^2 - (a + c)(b + d)p_1^*(1 - p_1^*)q_1^*(1 - q_1^*) = 0.$$

Since $0 < p_1^* < 1$ and $0 < q_1^* < 1$, the eigenvalues have opposite sign if $(a + c)(b + d) > 0$ and are pure imaginary if $(a + c)(b + d) < 0$. We have seen examples of both cases.

(4) In the second case, use the method of Subsection 9.4.2 to find a first integral. One answer:

$$V(p_1, q_1) = a \ln q_1 + c \ln(1 - q_1) - d \ln p_1 - b \ln(1 - p_1).$$

(5) Show that V has a local extremum at (p_1^*, q_1^*) by showing that $V_{p_1 p_1} V_{q_1 q_1} - V_{p_1 q_1}^2 > 0$ at (p_1^*, q_1^*). This implies that that near (p_1^*, q_1^*) the orbits of the replicator system are closed, so the solutions are time periodic. In fact all solutions in the interior of the rectangle $0 \leqslant p_1 \leqslant 1$, $0 \leqslant q_1 \leqslant 1$ are time periodic, but we will not show this.

Notice that when $(a+c)(b+d) \neq 0$, interior equilibria (p_1^*, q_1^*) are never asymptotically stable. This is a case of a more general fact: for *any* asymmetric game, interior equilibria are never asymptotically stable [8].

(6) Show that on any periodic solution, the average of $p_1(t)$ is p_1^*, and the average of $q_1(t)$ is q_1^*. (See the end of Section 10.5.)

Appendix

Sources for examples and problems

Chapter 1

Section 1.5. [4], sec. 1.2.
Section 1.8. [4], sec. 2.18. For other versions, see the Wikipedia page devoted to this game, http://en.wikipedia.org/wiki/Centipede_game.
Section 1.13. For backward induction in mixed martial arts, see [13].
Problem 1.14.3. [5], sec. 4.7.
Problem 1.14.4. [2], sec. 6.2.
Problem 1.14.5. [14].

Chapter 2

Section 2.6. [5], sec. 4.9.
Section 2.7. The notation in this problem comes from [5], sec. 4.6.
Section 2.9. [7].
Problem 2.14.2. [5], sec. 4.6.
Problem 2.14.3 (1). [5], sec. 4.3.
Problem 2.14.4. [5], sec. 4.10.
Problem 2.14.5. [5], sec. 4.15.
Problem 2.14.6. [6], sec. 4.11.
Problem 2.14.7. [6], sec. 4.10.

Chapter 3

Section 3.3. [5], sec. 6.5.
Section 3.5. [5], sec. 5.4.
Problem 3.12.1. [5], sec. 5.1.
Problem 3.12.2. [5], sec. 5.1.
Problem 3.12.3. [5], sec. 5.2 (a).
Problem 3.12.4. [5], sec. 5.2 (b).

Problem 3.12.7. [6], sec. 4.5.
Problem 3.12.9. [5], sec. 5.15 (a).
Problem 3.12.10. [5], sec. 4.12.
Problem 3.12.14. http://en.wikipedia.org/wiki/Braess's_paradox.
Problem 3.12.15. [9].

Chapter 4

Section 4.2. [4], sec. 3.18.
Section 4.4. [5], sec. 3.21.
Section 4.6. [3], chapter 14.
Problem 4.7.1. [3], sec. 3.7.

Chapter 5

Section 5.1. [5], sec. 6.4.
Section 5.4. [5], sec. 6.21.
Section 5.6. [1], p. 48.
Section 5.7. [5], sec. 6.29.
Problem 5.12.1. [5], sec. 6.10.
Problem 5.12.3. [5], sec. 6.14.
Problem 5.12.5. [5], sec. 6.3.
Problem 5.12.8. [5], sec.6.7.
Problem 5.12.10. [5], sec. 6.30.

Chapter 6

Section 6.5. [5], sec. 5.10.
Section 6.10. This example is a simplification of the version in [5], sec. 5.16.
Section 6.11. The setup for this example is from [5], sec. 5.13.
Problem 6.12.3. [5], sec. 5.14.
Problem 6.12.4. [5], sec. 9.4.

Chapter 7

Section 7.2. [12], sec. 4.8.
Section 7.3. [5], sec. 6.26.
Problems 7.5.2–7.5.3. [5], secs. 6.35 to 6.37.

Chapter 8

Section 8.4.3. [5], sec. 3.10.
Section 8.5. [5], sec. 6.26.
Problem 8.6.1. [5], sec. 10.9.
Problem 8.6.2. [5], sec. 6.2.
Problem 8.6.4. [5], sec. 6.41.

Chapter 10

Section 10.5. [5], secs. 6.25 and 12.14.
Problem 10.13.1. [5], sec. 6.2.
Problem 10.13.4. [5], sec. 12.13.
Problem 10.13.6. [5], sec 12.19.
Problems 10.13.7–10.13.8. [15].

References

[1] K. Binmore, *Game Theory: A Very Short Introduction*, Oxford University Press, New York, 2007.

[2] S. J. Brams, *Game Theory and the Humanities: Bridging Two Worlds*, MIT Press, Boston, 2012.

[3] A. K. Dixit and S. Skeath, *Games of Strategy*, second edition, Norton, New York, 2004.

[4] H. Gintis, *Game Theory Evolving*, first edition, Princeton University Press, Princeton, NJ, 2000.

[5] H. Gintis, *Game Theory Evolving*, second edition, Princeton University Press, Princeton, NJ, 2009.

[6] H. Gintis, *The Bounds of Reason: Game Theory and the Unification of the Behavioral Sciences*, Princeton University Press, Princeton, NJ, 2009.

[7] O. G. Haywood Jr., "Military decision and game theory," *Journal of the Operations Research Society of America* **2** (1954), 365–385.

[8] J. Hofbauer and K. Sigmund, *Evolutionary Games and Population Dynamics*, Cambridge University Press, Cambridge, 1998.

[9] L. R. Iannaccone, C. E. Haight, and J. Rubin, "Lessons from Delphi: Religious markets and spiritual capitals," *Journal of Economic Behavior and Organization* **72** (2011), 326–338.

[10] M. A. R. Kleiman, *When Brute Force Fails: How to Have Less Crime and Less Punishment*, Princeton University Press, Princeton, NJ, 2009.

[11] J. Nash, "Non-cooperative games," *Annals of Mathematics* **54** (1951), 286–295.

[12] M. Osborne, *An Introduction to Game Theory*, Oxford University Press, New York, 2004.

[13] M. Shaer, "Cage match: How science is transforming the sport of MMA fighting," http://www.popsci.com/science/article/2012-07/cage-match.

[14] I. Stewart, "A puzzle for pirates," *Scientific American* (May 1999), 98–99.

[15] K. Sigmund, C. Hauert, and M. Nowak, "Reward and punishment," *Proceedings of the National Academy of Sciences USA* **98** (2001), 10757–10762.

Index